A SOURCEBOOK

OF

PROBLEMS FOR GEOMETRY

BASED UPON

INDUSTRIAL DESIGN AND ARCHITECTURAL ORNAMENT

BY

MABEL SYKES

INSTRUCTOR IN MATHEMATICS IN THE BOWEN HIGH SCHOOL
CHICAGO, ILLINOIS

WITH THE COOPERATION OF

H. E. SLAUGHT AND N. J. LENNES

———◆◆◆◆◆———

DALE SEYMOUR PUBLICATIONS

Project Editor: Joan Gideon
Production Manager: Janet Yearian
Production Coordinator: Claire Flaherty
Design Manager: Jeff Kelly

The cover, designed by David Woods, is based on the drawings found on page 222.

This book is published by Dale Seymour Publications, an imprint of the Alternative Publishing Group of Addison-Wesley Publishing Company.

Originally published in 1912 by Norwood Press, Norwood, Mass.

Order number 21327
ISBN 0-86651-795-2

1 2 3 4 5 6 7 8 9 10-ML-97 96 95 94 93

DALE
SEYMOUR
PUBLICATIONS
P.O. BOX 10888
PALO ALTO, CA 94303

This Book Is Printed
on Recycled Paper

PREFACE TO REPRINT

MY fascination with old mathematics books began when I was a teacher and has continued during my tenure as publisher. One of my favorites has always been Mabel Sykes' *Source Book of Problems for Geometry*. Webster defines a classic as "a work of enduring excellence." As you become familiar with *Source Book of Problems for Geometry*, you will see that it has lasting quality.

- It contains unique design analysis of many patterns. Mabel Sykes points out in her preface that she gathered material of many geometric designs to help students develop a sense of form.

- It allows us to compare teaching approaches used at the beginning of the century to those used today. Though the approaches have changed, some of the attitudes are remarkably contemporary. The author's comment that "at present there is a widespread tendency in education to substitute amusement for downright work" may sound familiar to math teachers eighty years later.

- It provides a role model of a woman mathematician. Mabel Sykes wrote this book during a time when attitudes dictated that women had little business being involved with mathematics in any capacity.

Mabel Sykes used designs, such as the one on the cover, as a starting point for a proof, or as a basis for finding an area or an arc length. You and your students will find many ways to use and learn from the work of this industrious mathematician. You might select some problems as alternatives to those in your textbook (the Index to Problems and Theorems will help you find appropriate problems). Or use a few problems for enrichment or as a starting point for student investigations. You might also have students

- Analyze one of the designs.

- Construct one of the mathematical drawings.

- Work with a group to do all of the activities for one design and present the analysis and solutions to the class.

- Compare two theorems that could be proved from a mathematical drawing.

- Find a design that is not in the book, construct the corresponding mathematical drawing, and present theorems or problem based on the drawing.

- Use resources listed in the updated bibliography (included in place of the original outdated bibliography) to research geometry or the history of mathematics.

As an exciting investigation, invite students to conduct original historical research by having them find out more about Mabel Sykes. If your students uncover anything interesting about her work or her life, please let us know so we can share the information with others.

Dale Seymour

PREFACE

THE author is fully aware of the danger involved in offering this book to teachers of geometry.

At present there is a widespread tendency in education to substitute amusement for downright work. This temptation presents itself in varied and subtle forms to all teachers. Even the best teacher cannot hope to escape. He must be continually on his guard lest his work call forth mere superficial attention and begin and end in entertainment.

One form of this temptation may be found in the copying of designs. Not only may one spend more time than is profitable on this work, but unless care is exercised in the selection of the designs, all of the time spent may be wasted. Many interesting figures that might be suggested for this purpose are so complicated that pupils cannot reasonably be expected to determine all of the lines by constructions already mastered.

If, attracted by the beauty of the copy, the pupil is allowed to present a result that will not bear detailed geometrical analysis, entertainment has been substituted for education. *A design which is not an easily recognized application of a known construction is certainly of doubtful value.* Many of the designs in this book are very simple. A few are complicated, but even these can be so broken up that only the simple part need be given to the class.

While it must remain true that for training in clear thinking and accurate expression nothing in the teaching of geometry can take the place of abstract originals and of formal proof of theorems, nevertheless, exercises of the type here given have a unique and important place.

We no longer teach geometry solely because it develops the reasoning powers or trains in exact thought and expression. Such development and training can be obtained from other subjects rightly taught. But geometry gives, as no other subject can give, *an appreciation of form as it exists in the material world, and of the dependence of one form upon another.* This appreciation is apart from and superior to any particular utilitarian value that the subject may or may not have. Surely, one of the strongest reasons for teaching geometry in the high school is that it develops in pupils a sense of form. To cultivate and develop this sense in the highest degree, pupils should learn to recognize applications of geometric form which are omnipresent. The figures in the text-books which may have seemed unattractive and devoid of human interest, will then appear not only useful, but even beautiful. It is for this purpose that these exercises are of value.

This book is written in the hope that it may meet a real need of both teachers and pupils. Hitherto much material from geometric designs or historic ornament suitable for class use could be obtained only by long and careful search through widely scattered sources. The difficulties of this search and the labor of preparing the results have been such that a valuable and interesting element in the study of geometry has been for the most part lost until very recently.

Much time has been spent in collecting these illustrations. Moreover, care has been taken to select not only many simple designs, but also illustrations of simple geometric figures, inasmuch as these are in every way better suited to the purposes mentioned above than are complicated ones. A simple design is not necessarily a poor one. The references inserted amply justify the claim that these illustrations have their origin in what is best in historic ornament.

Reasonable limits forbid the insertion of many problems that are special cases under the same type. Instead, the general problem with its answer is given from which any number of numerical problems may be obtained.

The problems that combine both algebra and geometry are sure to be of value in bringing out the essential unity of the two subjects.

Wherever data for theorems and problems appear to be insufficient, what is lacking may be readily supplied from the figure. Attention is called to the index of theorems and problems as an aid to selecting material.

Without the assistance and coöperation of Dr. H. E. Slaught, of the University of Chicago, and Dr. N. J. Lennes, of Columbia University, this book could never have gone to press. The author wishes to express her deep appreciation of their continual encouragement and of the hours spent by each of them in detailed work on both manuscript and proof.

Thanks are due to Mr. Dwight H. Perkins, the Chicago architect, for criticisms and suggestions regarding the notes on trusses and arches, and also to many former pupils and fellow teachers who have assisted in verifying answers.

M. S.

CHICAGO, ILLINOIS, November, 1912.

TABLE OF CONTENTS

Chapter I.   Tile Designs

Chapter II.   Parquet Floor Designs

Chapter III.   Miscellaneous Industries

Chapter IV.   Gothic Tracery: Forms in Circles

Chapter V.   Gothic Tracery: Pointed Forms

TABLE OF CONTENTS

Chapter VI. Trusses and Arches

SOURCE BOOK OF PROBLEMS FOR GEOMETRY

CHAPTER I

TILE DESIGNS

PART 1. SIZES AND SHAPES OF STANDARD TILES

1. Occurrence. — Modern standard tiles are made in certain definite sizes and shapes, all of which are directly derived from the six-inch square. Some of the more common shapes and their relations to the square are shown in this Part. When tiles are intended for floor use, the pattern is worked out with plane unglazed tiles of a large variety of colors and shapes. Tiles intended for the walls of kitchens or bathrooms or for fireplaces are usually rectangular and often highly decorated. Among them are some of the most artistic products of modern industrial art.

2. History. — The earliest tiles of which we have any record are from the valleys of the Tigris, Euphrates, and Nile rivers.

The most famous examples of **tiled pavements** are those which occur in the churches and abbeys erected during the middle ages throughout western Europe. The tiles were mostly square, and highly ornamented. Reference will be made to some of the designs later.

Tiles for wall decoration were characteristic of the work of the Moslems of the eleventh and twelfth centuries, when they brought the art to great perfection. The tiles used were elaborately decorated and made in a large variety of shapes.

SQUARES, RECTANGLES, AND RIGHT TRIANGLES

3. In Fig. 1 ABCD represents a standard tile six inches square. EG and FH are the diameters of the square and AC and BD the diagonals. The points are joined as indicated.

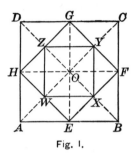

Fig. I.

EXERCISES

1. Prove that $EFGH$ and $XYZW$ are squares.

2. Find the length of one side of $EFGH$ and its area.

Ans. 4.242 in., 18 sq. in.

3. Find the length of one side of $XYZW$ and its area.

Ans. 3 in., 9 sq. in.

4. How many squares of the size of $XYZW$ would be sold for one square yard? *Ans.* 144.

4. In Fig. 2 ABCD represents a standard tile six inches square. $AE = AG$ and EF and FG are parallel to AG and AE respectively.

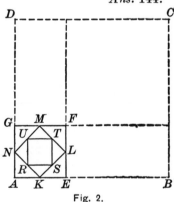

Fig. 2.

EXERCISES

1. Prove that $AEFG$ is a square.

2. Square $KLMN$ is formed by joining middle points of sides of $EFGA$ as in Fig. 1. Find one side of square $KLMN$ and its area if
(1) $AE = \frac{1}{2} AB$;   (2) $AE = \frac{1}{4} AB$;   (3) $AE = \frac{1}{3} AB$.

Ans. to (3), 2.12 in., $4\frac{1}{2}$ sq. in.

3. If $AE = \frac{1}{3} AB$, how many tiles of size of $KLMN$ would be sold for one square yard? *Ans.* 648.

4. How many whole squares of size of $KLMN$ could be cut from one square yard if $AE = \frac{1}{4} AB$? *Ans.* 1152.

5. Square $RSTU$ is formed by joining the middle points of sides of square $KLMN$, as in Fig. 1. Find one side of square $RSTU$ and its area if

    (1) $AE = \frac{1}{3} AB$;   (2) $AE = \frac{1}{4} AB$;   (3) $AE = \frac{1}{2} AB$.

Ans. to (3), $1\frac{1}{2}$ in., $2\frac{1}{4}$ sq. in.

6. If $AE = \frac{1}{4} AB$, how many tiles of size of $RSTU$ can be cut from one square yard? *Ans.* 2304.

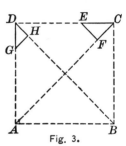

5. In Fig. 3 ABCD is a **six-inch square.** CE $= \frac{1}{3}$ CD and DG $= \frac{1}{4}$ DA.

Fig. 3.

EXERCISES

1. Show how to construct the isosceles right triangles with CE and DG as hypothenuses.

2. Find CF and DH and the areas of the triangles CEF and DGH. *Ans.* Sides, $\sqrt{2}$, $\frac{3}{4}\sqrt{2}$; Areas, 1, $\frac{9}{16}$.

3. If CF is $\frac{3}{2}\sqrt{2}$ in., find the ratio of CE to CD. *Ans.* $\frac{1}{2}$.

4. If $CF = a$, find the ratio of CE to CD. *Ans.* $\dfrac{a\sqrt{2}}{6}$.

5. If CE is $\dfrac{CD}{n}$, find the length of CF. *Ans.* $\dfrac{3}{n}\sqrt{2}$.

6. If $CE = \frac{1}{3} CD$ and $DG = \frac{1}{4} DA$, how many tiles of each size would be sold for one square yard? *Ans.* 1296 and 2304.

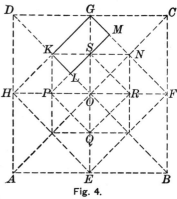

Fig. 4.

6. In Fig. 4 ABCD represents a six-inch square. E, F, G, and H are the middle points of the sides. The points are joined as indicated.

EXERCISES

1. Prove that

(a) $KL = LO$ and PS extended will bisect GN;

(b) $GKLM$ is a rectangle;

(c) Square $GKON \cong PQRS$, and in area equals $\frac{1}{2} \triangle DOC$;

(d) The area of rectangle $GKLM$ is equal to that of $\triangle DKG$ and to $\frac{1}{4}$ that of $\triangle DOC$.

2. How many rectangles of the size of $GKLM$ can be cut from one square yard?   *Ans.* 576.

3. Find the dimensions of rectangle $GKLM$.   *Ans.* 2.121, 1.06.

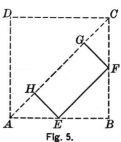

Fig. 5.

7. In Fig. 5 ABCD represents a six-inch square. E and F are respectively the middle points of the sides AB and BC. EH and FG are drawn from E and F respectively perpendicular to AC.

EXERCISES

1. Prove that $EFGH$ is a rectangle.

2. Prove that the area of $EFGH$ is equal to: (a) twice that of $\triangle EBF$; (b) half that of $\triangle ABC$; (c) four times that of $\triangle AEH$; (d) half that of the square formed by joining the middle points of sides of square $ABCD$.

3. Find the dimensions of rectangle $EFGH$, and the number that make up one square yard.   *Ans.* (1) 4.242 × 2.121; (2) 144.

IRREGULAR PENTAGONS, HEXAGONS, AND OCTAGONS

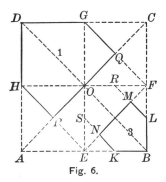
Fig. 6.

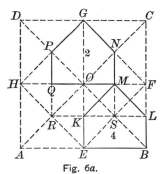
Fig. 6a.

8. Irregular pentagons. — In Figs. 6 and 6a ABCD represents a six-inch square. E, F, G, and H are the middle points of the sides. In Fig. 6 K, L, R, and S are the middle points of the sides of the square EBFO. In both figures the points are joined as indicated to form the irregular pentagons numbered 1, 2, 3, and 4.

EXERCISES

1. In Fig. 6 find the length of *HP* and *PQ*, and the area of pentagon 1.

Ans. (1) 2.121 in.; (2) 4.242 in.; (3) $13\frac{1}{2}$ sq. in.

2. In Fig. 6 find the length of *KN* and *MN* and the area of pentagon 3.

3. In Fig. 6a find *PQ*, *PG*, and the area of pentagon 2.

Ans. (1) 1.5 in.; (2) 2.121 in.; (3) $6\frac{3}{4}$ sq. in.

4. In Fig. 6a prove that *RS* extended bisects *OE* and *FB*.

5. If the points *E*, *B*, *L*, *M*, and *K* (Fig. 6a) are joined as indicated, prove that pentagon 2 is congruent to pentagon 4.

6. Prove that pentagons 1 and 3 in Fig. 6 and pentagon 2, Fig. 6a, are similar. Show that their areas obey the law for the ratios of the areas of similar polygons.

7. Show how to construct from the square *ABCD* a pentagon similar to pentagon 3, but having one half its area. Show how to construct one similar to pentagon 2, but having one half its area.

Suggestion. — See Ex. 6 above.

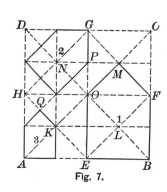

Fig. 7.

Fig. 7a.   Tiled Flooring.
Scale ¾ in.=1 ft.

9. In Fig. 7 ABCD represents the six-inch square, and E, F, G, and H are the middle points of the sides. The points are joined as indicated.

EXERCISES

1. Prove that the line joining N and K bisects AE, HO, and DG.

2. If the three pentagons are formed as indicated, prove them similar.

3. Find (a) the ratio of the sides FM, GP, and KQ, and (b) the ratio of the areas of pentagons 1, 2, and 3.

$$Ans.\ (a)\ \sqrt{2}:1:\frac{1}{\sqrt{2}};\ (b)\ 2:1:\tfrac{1}{2}.$$

4. Find the areas of pentagons 1, 2, and 3.

$$Ans.\ 11\tfrac{1}{4}\ \text{sq. in.},\ 5\tfrac{5}{8}\ \text{sq. in.},\ 2\tfrac{13}{16}\ \text{sq. in.}$$

5. Figure 7a contains two sizes of tiles that are irregular pentagons. Show the ratio between each and the six-inch square.

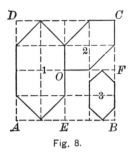

Fig. 8.

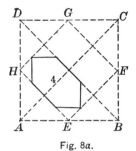

Fig. 8a.

10. Irregular hexagons. — In Fig. 8 and 8a ABCD represents the six-inch square. Points E, F, G, and H are the middle points of the sides. Each side of squares ABCD and EBFO in Fig. 8 and each side of EFGH in Fig. 8a is divided into four equal parts and the points joined to form the hexagons numbered 1, 2, 3, and 4.

EXERCISES

1. Prove that in each hexagon the opposite sides are parallel.

2. Find the length of the sides and the area of each hexagon.

Ans. for No. 4, 2.121 in., $1\frac{1}{2}$ in., $6\frac{3}{4}$ sq. in.

3. How many hexagons of each kind would be sold for one square yard? *Ans.* 96 of number 1.

4. Show that hexagons 2 and 4 are congruent.

5. Prove that hexagons 1, 2, 3, and 4 are similar, and that their areas obey the law for the ratio of the areas of similar polygons.

6. From Ex. 5, show how to construct a hexagon from square *ABCD*, similar to hexagon 1, but having twice its area.

7. Figure 8*b* shows two tiles that are irregular hexagons. Show the relation between each and the six-inch square.

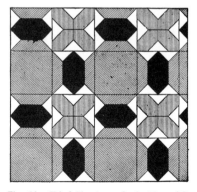

Fig. 8*b*. Tiled Flooring. Scale $\frac{3}{4}$ in. $=$ I ft.

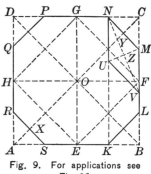

Fig. 9. For applications see Fig. 26.

11. Irregular octagons. — In Fig. 9 ABCD represents a six-inch square. Each side is divided into four equal parts and the points joined to form the octagon KLMN, etc.

EXERCISES

1. Find the ratio of *KL* to *LM* and of *GE* to *XY*.

$$Ans. \quad \frac{\sqrt{2}}{2} \text{ and } \frac{3\sqrt{2}}{4}.$$

2. Prove that *XY* is the perpendicular bisector of *MN*.

3. Find the area of the octagon.

$$Ans. \ 31^1 \text{ sq. in.}$$

4. If similar octagons were formed from squares *EFGH* and *EBFO*, what would be their areas?

$$Ans. \ 15\tfrac{3}{4} \text{ sq. in., } 7\tfrac{7}{8} \text{ sq. in.}$$

5. Prove that *EKLF* is a trapezoid and find its area.

$$Ans. \ 3\tfrac{3}{8} \text{ sq. in.}$$

6. Join *NK*. Prove that *NK* is parallel to *BC*. Complete the rhombus *MNUV*, and find its area.   *Ans.* 3.181 sq. in.

7. Find the lengths of the diagonals *NV* and *UM*.

$$Ans. \ \tfrac{3}{2}\sqrt{4 + 2\sqrt{2}}, \ \tfrac{3}{2}\sqrt{4 - 2\sqrt{2}}.$$

Suggestion. — Use △ *VNC* and *NZM* to find lengths of *NV* and *UM* respectively.

8. Find the area of square constructed on *CY* as diagonal.

$$Ans. \ \tfrac{9}{16} \text{ sq. in.}$$

Suggestion. — Use the theorem: "The bisector of an angle of a triangle divides the opposite side into segments proportional to the adjacent sides."

12. In Fig. 10 ABCD represents the six-inch square. E, F, G, H are the middle points of the sides. AP = QE = NO, etc. = ⅙ AE.

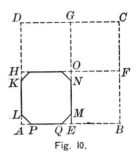

Fig. 10.

EXERCISES

1. Find the area of the octagon *PQMN* etc.   *Ans.* 8½ sq. in.

2. Find the ratio of the sides *PQ* and *QM*.   *Ans.* $2\sqrt{2}:1$.

13. In Fig. 11 ABCD represents the six-inch square. E, F, G, H are the middle points of the sides. L and M are the middle points of KF and FN respectively, and the points are joined to form the figure KLMNO. R and Q are the middle points of HX and XE respectively. RS and QP are parallel to HO and OE respectively, forming figure APQRS.

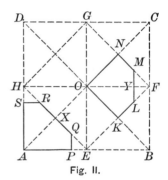

Fig. 11.

EXERCISES

1. Find the areas of *KLMNO* and *APQRS*.   *Ans.* 3 15/16 sq. in.

2. Construct a figure similar to *KLMNO* from octagon *KLMN* etc., Fig. 9, and find its area.   *Ans.* 7⅞ sq. in.

3. Show how to construct from square *ABCD* a figure similar to *APQRS*, but having twice its area.

Suggestion. — It is ¼ the octagon *KLMNP* etc., Fig. 9.

4. Show how to construct from square *ABCD* figures similar to *KLMNO* and *APQRS*, but having ½ their areas.

Suggestion. — Use the octagon derived from square *AEOH*.

5. Show that figures *MNOY* and *RSAX* are congruent. Are *KLMNO* and *APQRS* congruent?

REGULAR HEXAGONS AND RELATED FIGURES

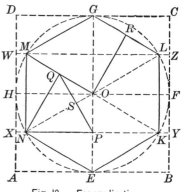

14. In Fig. 12 ABCD represents a six-inch square with its inscribed circle. Sides BC and DA are each divided into four equal parts and the points joined by lines WZ, HF, and XY. Lines WZ and XY cut the circle at points N, K, L, and M. Points Q and R are the middle points of OM and GL respectively.

Fig. 12 — For applications see Figs. 17–19.

EXERCISES

1. Show how to inscribe a circle in the given square, and prove that the points of tangency are the middle points of the sides.

2. Prove that *EKLGMN* is a regular hexagon, if *E* and *G* are the middle points of *AB* and *DC* respectively.

3. Prove *KL* parallel to *BC*.

4. Prove that *NPQ* is an equilateral triangle, and *ORGM* is a trapezoid.

5. Find the area of the circle and of the curved figure *EKFB*.
Ans. 28.27 sq. in., 1.93 sq. in.

6. How many circular tiles six inches in diameter would make one square yard?   *Ans.* 45.

7. Find the areas of (*a*) hexagon *EKLG* etc., (*b*) △ *PQN*, (*c*) △ *OPQ*.   *Ans.* $\dfrac{27\sqrt{3}}{2}$, $\dfrac{27\sqrt{3}}{16}$, $\dfrac{9\sqrt{3}}{16}$.

Suggestion for (*c*).—Show that *OS* is perpendicular to *PQ* and that $OS = \frac{3}{4}$ in.

8. What would be the length of one side of the circumscribed square, if the area of the hexagon is (*a*) $\frac{1}{2}$ that of hexagon *EKLG* etc., (*b*) $\frac{1}{4}$ that of hexagon *EKLG* etc.   *Ans.* for (*b*), 3.

9. Construct each of the hexagons mentioned in Ex. 8 from a six-inch square.

Suggestion. — They must be constructed in squares *EFGH* and *AEOH* respectively.

10. Find the area of trapezoid *ORGM* (1) by the formula for the area of the trapezoid, (2) by showing that it is ¼ of the hexagon *EKLG* etc. *Ans.* $\frac{27}{8}\sqrt{3}$.

11. How may a figure similar to *ORGM*, but having ½ its area, be constructed from a six-inch square?

Suggestion. — Use square *EFGH*.

12. How may a figure similar to *ORGM*, but having ¼ its area, be constructed from a six-inch square?

CIRCULAR FORMS

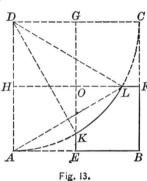

15. In Fig. 13 ABCD represents the six-inch square. EG and FH are its diameters. Arc AKLC is drawn with D as center and DA as radius. Lines DK, DL, and AL are drawn.

Fig. 13.

EXERCISES

1. Prove that $\angle ADK = \angle KDL = \angle LDC = 30°$.

Suggestion. — Prove *ALD* an equilateral △.

2. Find the length of *HL* and the area of △ *HLD*.

Ans. (1) $3\sqrt{3}$ in.; (2) $\frac{9}{2}\sqrt{3}$ sq. in.

3. Find the area of the sector *LDC*. *Ans.* 9.4248 sq. in.

4. Find the area of the figure *LFC*.

Ans. $\frac{3}{2}(12 - 3\sqrt{3} - 2\pi)$.

Suggestion. — Subtract the sum of the areas of the sector *LDC*, and the △ *LDH* from the rectangle *FCDH*.

5. Find the area of *EBFLK*.

Suggestion. — Subtract the sum of the areas of *FCL* and *EKA* and the sector *CDA* from the area of the square.

Ans. 6.16⁺ or $3(3\sqrt{3} - \pi)$.

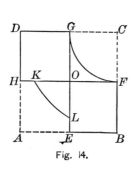

Fig. 14.

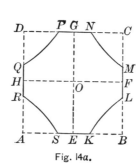

Fig. 14*a*.

Fig. 14*b*.   Tiled Flooring.
Scale $\frac{3}{8}$ in. = 1 ft.

16. In Figs. 14 and 14*a* ABCD represents the six-inch square with the diameters EG and FH. In Fig. 14 $\overset{\frown}{GF}$ and $\overset{\frown}{KL}$ are drawn with C as center and CG and CD respectively as radii. In Fig. 14*a* AB is the radius for $\overset{\frown}{KL}$, $\overset{\frown}{MN}$, $\overset{\frown}{PQ}$, and $\overset{\frown}{RS}$.

EXERCISES

1. If *SEOHR* in Fig. 14*a* is congruent to *AELKH* in Fig. 14, find the center for $\overset{\frown}{RS}$.

2. Show that *AELKH* in Fig. 14 is congruent to *EBFLK* in Fig. 13, if the figures are on the same scale.

3. Find the area of *EBFGDHKL* in Fig. 14 and of *KLMNP* etc. in Fig. 14*a*.   *Ans.*   (1) 22.77 sq. in.; (2) 24.64 sq. in.

17. In Fig. 15 ABCD represents the six-inch square, with EG and FH the diameters. Arc KOL is drawn with B as center and the half diagonal BO as radius.

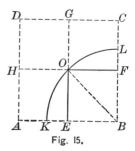

Fig. 15.

1. Find the area of the figure *FOL*.   *Ans.* $\frac{9}{4}(\pi - 2)$.

Suggestion. — Show that the sector $OBL = \frac{1}{8}$ of the circle.

2. Find the length of *FL*.   *Ans.* $3(\sqrt{2} - 1)$.

18. In Fig. 16 ABCD represents the six-inch square with EG one diameter.

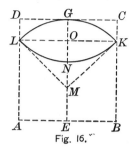

Fig. 16.

1. Show how to construct the arc *KGL* tangent to *CD* at *G*, if its radius is $\frac{1}{2}$ the diagonal of the square.

2. Show how to construct the arc *LNK* so that it shall have the same radius as the arc *KGL* and shall pass through the points *K* and *L*.

3. Prove that $\angle LMK$ is 90°.

Suggestion. — Since $MK = 3\sqrt{2}$ and $KO = 3$, $OM = 3$, then $\angle OMK = 45°$.

4. Find the area of the oval *KGLN*.   *Ans.* $9(\pi - 2)$.

Suggestion. — Find the area of the segment *KGL* from the areas of the sector *MKGL* and the $\triangle MKL$.

5. Find the length of *GN*.   *Ans.* $6(\sqrt{2} - 1)$.

6. If arc *KGL* is a quadrant and is constructed tangent to *CD* at *G*, prove that its radius is $\frac{1}{2}$ the diagonal of the square.

PART 2. TILED FLOOR PATTERNS

19. Designing in systems of parallel lines. — Many repeating patterns commonly used in industrial designs are planned on one or more systems of parallel lines. When the design is to be in stripes, one system of parallels is necessary. (See Fig. 112.) When it is based on parallelograms, two systems are needed. Rhombuses are often used for wall papers and fabrics, and squares for tiled or parquet floors and linoleums. When the design is based on triangles or hexagons, three systems of parallels are used. Such designs are freely used for tiled floors. They were highly characteristic of Arabic ornament. When octagons are desired, four systems of parallels must be drawn. Such designs are frequently used for linoleums and steel ceilings as well as for tiled floors.

20. Designing tiled floors. — Tiled floor designs are planned on systems of parallel lines. Moreover, modern tiles are made in certain definite sizes and shapes which are related to each other as shown in the previous section. In planning tiled floor designs, therefore, the designer so regulates the distances between the parallels and the angles at which they intersect as to determine the tiles which he wishes to use. This is evident from a study of the designs here shown.

21. History. — The patterns used for tiled floors are among the simplest found in industrial art. These simple designs abound in the geometric mosaics of southern and eastern Europe.[1] Italy and Sicily are full of famous examples, most of which are of Byzantine origin, although those in Sicily strongly suggest Moslem workmanship. There were many different kinds that received special names; among them the *Opus Alexandrinum* was an arrangement of small cubes of porphyry and serpentine set in grooves of white marble.

---

[1] See Gravina and Ongania for sample pavements.
See also J. Ward, (1), Vol. I, pp. 288–294, Vol. II, p. 344.

DESIGNS BASED ON THREE SYSTEMS OF PARALLEL LINES

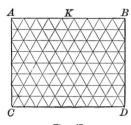

Fig. 17.

Fig. 17a. — Scale ¾ in. = 1 ft.

22. Figure 17 shows a pattern involving equilateral triangles constructed by drawing three sets of parallel lines.

EXERCISES

1. At what angle must the lines intersect?

2. If Fig. 17a is drawn on a scale of ¾ in. to 1 ft., find the length of each side of one triangle. *Ans.* 3 in.

3. How may one of the triangles shown in Fig. 17a be derived from the six-inch square?

Suggestion. — See Fig. 12.

4. If a side of each triangle is 3, find the area of each.

Ans. $\frac{9}{4}\sqrt{3}$.

5. If a side of each triangle is $\frac{3}{2}\sqrt{2}$, find the area of each.

Ans. $\frac{9}{8}\sqrt{3}$.

6. Construct the pattern shown in Fig. 17a, using if possible drawing board, T square, and triangle. Let one side of each triangle represent 3 inches. Scale 3 in. = 1 ft.

7. What will be the actual number of square inches of drawing paper, covered by one triangle, if the drawing is made as suggested in Ex. 6? *Ans.* $\frac{9}{64}\sqrt{3}$.

8. Parallel lines may be constructed by using T square and triangle. On what theorems may the construction be based?

Fig. 18. — Parquet Flooring.

Fig. 18*a.* — All-over Lace Pattern.

EXERCISES

23. **1.** If each side of a hexagon shown in Fig. 18 represents 3 inches, measure the accompanying cut and find the scale to which it is drawn.

2. Construct Fig. 18 from three systems of parallel lines.

3. If each side of a hexagon shown in the figure represents 3 inches, what is the distance represented between the oblique parallels?

4. Construct the design shown in Fig. 18 from an all-over pattern of equilateral triangles. Let each side of hexagon represent 3 inches. Scale $\frac{7}{8}$ in. = 1 ft.

Fig. 19. — Tiled Flooring. Scale $\frac{5}{8}$ in. = 1 ft.

EXERCISES

24. **1.** Construct the three sets of parallel lines on which Fig. 19 is designed. Is the pattern of equilateral triangles necessary?

2. If one side of the hexagon is 3, find the area of the star formed of one hexagon and six triangles. *Ans.* $27\sqrt{3}$.

3. Prove that if the sides of a regular hexagon are extended until they intersect, six equilateral triangles are formed whose combined area is equal to the area of the given hexagon.

4. Construct the design shown in Fig. 19, letting a side of the hexagon represent 3 inches. Scale 2 in. = 1 ft.

5. If the drawing is made as suggested in Ex. 4, what will be the actual number of square inches of drawing paper included in the star? *Ans.* $\frac{3}{4}\sqrt{3}$.

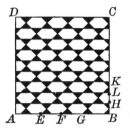

Fig. 20.—Tiled Flooring. Scale $\frac{3}{8}$ in. = 1 ft.

EXERCISES

25. 1. Construct the three systems of parallel lines on which Fig. 20 is based.

Suggestion. — The oblique lines form a system of squares. The horizontal lines bisect the sides of the squares.

2. Find the ratio (1) between a side of one of the squares and *KL*, and (2) between *EF* and *KL*. *Ans.* (1) $\dfrac{2}{\sqrt{2}}$; (2) 2.

3. If the scale of the figure is $\frac{3}{8}$ in. = 1 ft., find (1) the true length of *EF*; (2) the length of one side of one square; and (3) the sides of the triangle and of the irregular hexagons.

Ans. (1) 6; (2) $3\sqrt{2}$; (3) 3 and $\frac{3}{2}\sqrt{2}$.

4. Construct a design like Fig. 20. Let *EF* represent 3 inches. Scale 4 in. to 1 ft.

DESIGNS BASED ON FOUR SYSTEMS OF PARALLEL LINES

26. Irregular octogons.
— Figure 21 is constructed on four sets of parallel lines, as shown.

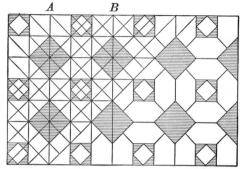

Fig. 21. — Tiled Flooring.   Scale $\frac{3}{4}$ in. = 1 ft.

EXERCISES

1. Find AB if one side of the larger dark square is $4\frac{1}{4}$ inches.

Ans. 9.01+.

2. If AB is exactly 9 inches, what is the length of one side of the larger square? *Ans.* $3\sqrt{2}$.

3. Construct the design for Fig. 21. Let each side of the large square represent 3 inches. Scale 4 in. to 1 ft.

4. If the drawing is made as in Ex. 3, how much actual paper is contained in the smallest square and in the irregular hexagon?

Ans. (1) $\frac{1}{4}$ sq. in; (2) $\frac{3}{4}$ sq. in.

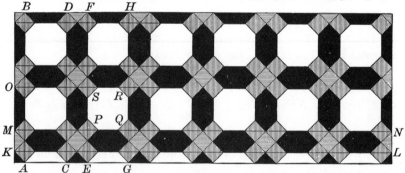

Fig. 22. — Tiled Flooring.   Moorish.   Scale $\frac{3}{4}$ in. = 1 ft.   Calvert (I), p. 301.

27. Figure 22 is based on four sets of parallel lines. $AC = 2\,CE = EG$, etc. $MO = 2\,MK$. Each side of the larger square is divided into four equal parts and diagonal lines are drawn to form the crosses and octagons.

EXERCISES

1. Construct a design like that shown in Fig. 22. Let AC represent 6 inches. Scale 2 in. = 1 ft.

2. If $AC = 6$, find the area of (1) the octagon ; (2) the irregular hexagon ; (3) one of the small squares.

$Ans.$ (1) $31\frac{1}{2}$; (2) $13\frac{1}{2}$; (3) $4\frac{1}{2}$.

3. If the drawing is made as suggested in Ex. 1, find the actual number of square inches of paper covered by each of the figures mentioned in Ex. 2. $Ans.$ (1) $\frac{7}{8}$; (2) $\frac{3}{8}$; (3) $\frac{1}{8}$.

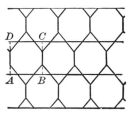

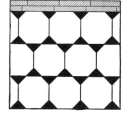

Fig. 23.   Fig. 23*a.* — Tiled Flooring.   Scale $\frac{9}{16}$ in. = I ft.

28. In Fig. 23 the horizontal and vertical sides of the octagons are equal, and the triangles included are isosceles right triangles.

EXERCISES

1. Find the ratio between the sides of the octagon. $Ans.$ $1 : \sqrt{2}$.

2. Show how the octagon may be constructed from a square.

3. Show how the design may be constructed from four sets of parallel lines.

Suggestion. — How does the distance between the vertical parallels compare with the distance between the horizontal ones?

4. Make a drawing to scale for the pattern shown in Fig. 23. Let the vertical side of the octagon represent 3 inches. Scale 3 in. = 1 ft.

5. In the drawing made for Ex. 4 how much actual paper is there (1) in one of the octagons and (2) in one of the triangles?

$Ans.$ (1) $\frac{63}{32}$ sq. in. ; (2) $\frac{9}{64}$ sq. in.

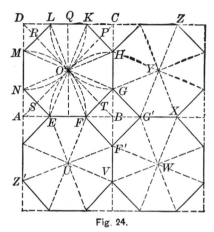

Fig. 24.

Fig. 24a. — Tiled Flooring.   Scale ⅝ in. = 1 ft.

29. Regular octagons. — In Figs. 24 and 24*a* the octagons shown are regular.

EXERCISES

1. Show how to inscribe a regular octagon in a square.

Suggestion. — *First Method :* With the vertices of square $ABCD$ as centers and the half diagonal OA as radius draw arcs cutting the sides of the square at points E, F, G, H, etc. To prove $EFGH$ etc. a regular octagon, show that its sides and angles are equal. Use ▲ EOF, FOG, GOH, etc. Notice that ▲ AOF, OBE, etc. are congruent and isosceles and that $\angle FOB$ is measured by ½ arc OF.

Second Method : With O as center and the half diameter OQ as radius cut the diagonals at points P, R, S, and T. Draw HK, LM, NE, and GF perpendicular to the diagonals through points P, R, S, and T. Now prove $EFGHKL$ etc. a regular octagon.

2. Find the ratios $AF:AB$, $AE:AB$, $AE:EF$.

$$Ans. \ \frac{\sqrt{2}}{2}, \ \frac{2-\sqrt{2}}{2}, \ \frac{\sqrt{2}}{2}.$$

3. If $AB = a$, find (1) the length of AE, (2) the area of $\triangle ANE$, and (3) the area of the octagon $EFGH$ etc.

$$Ans. \ (1) \ \frac{a}{2}(2-\sqrt{2}); \ (2) \ \frac{a^2}{4}(3-2\sqrt{2}); \ (3) \ 2 \, a^2(\sqrt{2}-1).$$

4. If the area of the octagon is 100 square inches, find the length of AB. $Ans. \ 5\sqrt{2\sqrt{2}+2}.$

5. If the area of the octagon is S, find the length of AB.

$$Ans. \; \tfrac{1}{2}\sqrt{S(2\sqrt{2}+2)}.$$

6. Construct the diagram for Fig. 24a. Let wx represent 6 inches. Scale 4 in. = 1 ft.

Suggestion. — See Fig. 24. Construct a network of squares. Inscribe a regular octagon in each square.

7. In Fig. 24 prove that $FGG'F'$ is a square.

8. Prove that $UVWXYHOE$ is a regular octagon.

9. Prove that $UFOE$ and $WXYG'$ are congruent rhombuses.

10. Prove that corresponding diagonals as EK and $G'Z$ are parallel.

11. Prove that points Z', F, and Z are collinear.

12. Construct Fig. 24a by drawing four sets of parallel lines and erasing those parts which are not necessary for the figure.

13. If in Fig. 24a $wx = a$, find the length of (1) CB; (2) AB; (3) the perpendicular distances between AD, BE, and CF.
Ans. (1) $a(2-\sqrt{2})$; (2) $a(\sqrt{2}-1)$; (3) $\frac{a}{2}(2-\sqrt{2})$ and $a(\sqrt{2}-1)$.

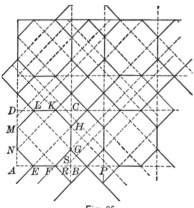

Fig. 25.

Fig. 25a. — Linoleum Pattern.

30. In Fig. 25 the octagons are regular, and the small quadrilaterals are squares.

EXERCISES

1. Find the ratio $\dfrac{BP}{AB} = \dfrac{2FB}{AB}.$
$Ans. \; \dfrac{2-\sqrt{2}}{1}.$

2. Construct a diagram for Fig. 25*a* by drawing four sets of parallel lines and erasing those parts of each which are not necessary in
the figure. Let $AB = 1\frac{1}{2}$ inches. Take $BP = 2\,BF$.

3. Prove that the construction suggested in Ex. 2 will make all the
octagons regular and all the small quadrilaterals squares.

4. If $AB = a$, find the length of FR.   *Ans.* $a(3 - 2\sqrt{2})$.
Suggestion. — Prove $HS = BP$; $GS = FR = HS - HG$;
$$HG = a(\sqrt{2}-1).$$

5. If $AB = a$, find the perpendicular distances between the oblique
parallels. *Ans.* (1) $a(\sqrt{2} - 1)$; (2) $\frac{a}{2}(2 - \sqrt{2})$; (3) $\frac{a}{2}(3\sqrt{2} - 4)$.

6. Can you suggest any other method for developing Fig. 25?
Suggestion. — See § 84.

31. In Fig. 26 the octagons are
regular.

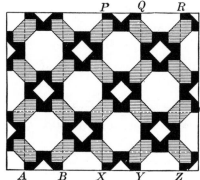

Fig. 26. — Linoleum Pattern.

EXERCISES

1. Determine the sets of parallel lines on which Fig. 26 may
have been constructed. Is there more than one possible method for
developing this figure?

Suggestion. — See § 85.

2. Find the relation between the long and the short side of the
irregular hexagon.   *Ans.* $2\sqrt{2} : 1$.

3. If $AB = a$, find the area of the cross, and of the small square.
       Ans. (1) $a^2(4\sqrt{2} - 5)$; (2) $a^2(3 - 2\sqrt{2})$.

CHAPTER II

PARQUET FLOOR DESIGNS

32. Parquet floors[1] are made by gluing or nailing into geometric patterns strips or blocks of various kinds of hard wood. These pieces of wood are cut with a scroll saw at the mills. In preparing the pattern, therefore, the designer is restricted to the use of pieces of such shapes as can be cut out conveniently. This accounts for the predominance of straight lines and for the simple geometric figures that enter into the combinations.

33. History. — Like tiled floor designs, most of the parquetry patterns are extremely simple, so that while suggestions for these designs seem to have been taken from every epoch and every country, only the simpler forms of historic ornament have been used.

Of the designs here given Figs. 39, 40, 55, 56, 64 are found in Medieval Byzantine Mosaics. Figures 48, 52, 55, 57, 60, 62 are of Arabic origin. Figure 51 is a closely related form from a Roman mosaic. Greek frets are also found in the catalogues. The simpler Mohammedan designs here referred to are from Cairo. Simple geometric patterns are also found in Chinese and Japanese art.

[1] A large proportion of the designs in this chapter are from parquet floor catalogues. See list of references.

PART 1. DESIGNS IN RECTANGLES AND PARALLELOGRAMS

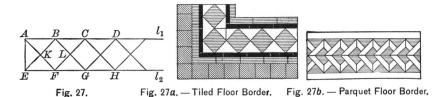

Fig. 27.   Fig. 27a. — Tiled Floor Border.   Fig. 27b. — Parquet Floor Border.

EXERCISES

34. **1**. Construct a series of squares in a given rectangle, as shown in Fig. 27.

Suggestion. — Divide the rectangle into squares by a series of parallels.

2. Prove that squares BF and CG are congruent.

3. How may the points BCD and FGH be laid off on the sides of the rectangle so that the inscribed squares may be constructed by drawing the straight lines EB, FC, AF, BG, etc.?

4. What relation must exist between two adjacent sides AE and AD of the rectangle in order that an integral number of squares may be inscribed in it?

5. If the border is 12 feet long, what must be its width in order that 16 squares may be exactly inscribed in it?   *Ans.* 9 in.

6. If the border on one side of a floor is 14 feet long, for what widths of border between 8 and 12 inches will it exactly contain an integral number of squares?

Ans. 8, 12, $10\frac{1}{2}$. If fractions are permissible, there is no limit to the possible number of results.

7. What is the length of a side of one of the inscribed squares if $AE = 2$? If $AE = a$?   *Ans.* $\sqrt{2}$, $\frac{a}{2}\sqrt{2}$.

8. Find the length and width of a border which contains 24 inscribed squares whose sides are each 8 inches.   *Ans.* $8\sqrt{2}$, $192\sqrt{2}$.

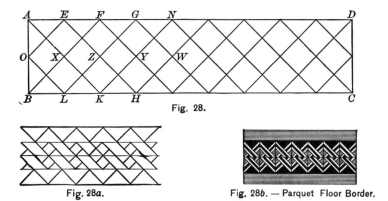

Fig. 28.

Fig. 28a.   Fig. 28b. — Parquet Floor Border.

EXERCISES

35. **1.** In Fig. 28 how are the points E, F, G, N and H, K, L laid off on the sides of the rectangle so that the lines BF, LG, EH, etc., will form two series of squares superimposed with the corner of one at the center of the next?

2. Prove that the lines BF, LG, etc., are parallel.

3. Prove that the lines BF and EH are perpendicular.

4. Prove that BF bisects EZ.

5. Prove that the vertices X and Y of the square $XKYF$ are the centers of the squares $OLZE$ and $ZHWG$.

6. Prove that the square $OEZL$ is divided into four equal squares.

7. What fraction of the square $XFYK$ does not form a part of any other square?   *Ans.* $\frac{1}{2}$.

8. Make a drawing for the border shown in Fig. 28a.

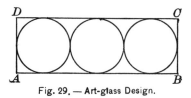

Fig. 29. — Art-glass Design.

EXERCISES

36. **1.** In Fig. 29 what must be the relation between the length and width of the rectangle $ABCD$ in order that the tangent circles may be inscribed as shown?

2. If in a rectangle three tangent circles are inscribed, what is the ratio of the area within the circles to the area outside the circles?

Ans. 3.66.

3. If in a rectangle three tangent circles are inscribed, what is the ratio of the area within the circles to the area of the whole rectangle?

Ans. .7854.

4. Does this ratio depend upon the shape and size of the rectangle?

5. Does the number of inscribed circles depend upon the size or shape of the rectangle or upon both ? Why?

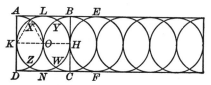

Fig. 30. — Art-glass Design.

EXERCISES

37. **1.** How must the points A, L, B, E, and D, N, C, F be laid off on the sides of the rectangle shown in Fig. 30 so that two series of tangent circles may be inscribed in the rectangle with the center of one circle on the circumference of the next?

2. If KH is the line of centers and O the point of tangency, prove that $\triangle KOX$ is equilateral.

3. If AD is 12 inches, find the area of the oval $KZOX$.

Ans. $6(4\pi - 3\sqrt{3})$.

4. If $KO = a$, find the area of the oval $KZOX$.

Suggestion. — Draw the chords KX and XO. The sector KOX is one sixth of the circle and its area is $\dfrac{\pi a^2}{6}$. The segment bounded by the chord KX and the arc KX is the difference between the area of an equilateral triangle whose side is a and the area of the sector KOX. The area of the segment is $\dfrac{\pi a^2}{6} - \dfrac{a^2}{4}\sqrt{3}$. Adding this to the area of the sector, we find the area of the half oval to be $\dfrac{a^2}{12}(4\pi - 3\sqrt{3})$.

5. If $AD = 12$, find the area of the figure $LXOY$.

Ans. $6(3\sqrt{3} - \pi)$, or 12.32+.

6. If $AD = a$, find the area of the figure $OYLX$.

$$Ans. \ \frac{a^2}{24}(3\sqrt{3} - \pi).$$

7. If the area of the oval $KXOZ$ is 49 square inches, find AD.

$$Ans. \ 12.6^+.$$

8. If the area of the figure $OYLX$ is s, find AD. $\quad Ans. \ \sqrt{\dfrac{24s}{3\sqrt{3} - \pi}}.$

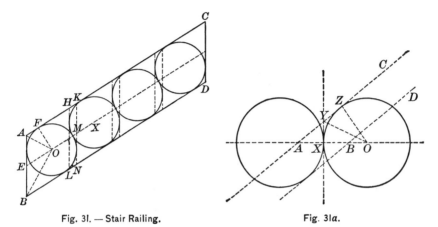

Fig. 3I. — Stair Railing.   Fig. 3Ia.

38. In Fig. 31 ABCD is a parallelogram with four tangent circles inscribed.

EXERCISES

1. If the lines AC and BD are supposed to be indefinite in extent, show how to construct circle O tangent to the lines AC, AB, and BD, and circle X tangent to lines AC and BD and to circle O.

2. If E is the middle point of AB and O and X are the centers of the circles, prove that the points E, O, and X are collinear.

3. Prove that $\angle AOB$ is 90° and that $OE = \frac{1}{2} AB$.

4. If $\angle ABL = 60°$, prove that $AO = \frac{1}{2} AB$, and that the radius of circle O is $\dfrac{AB}{4}\sqrt{3}.$

5. In Fig. 31a *A C* is tangent to circle *O* and *DB* to the other circle. *CA* and *DB* are parallel, and the circles are tangent at point *X*. Prove that *A B* is bisected at point *X*. The two circles are equal.

Suggestion. — Use one pair of congruent triangles. An additional construction will be necessary.

6. In Fig. 31a if ∠ *CAO* is 30°, show that *A C* passes through the center of the left circle.

7. Construct a figure like Fig. 31a, making ∠ *CAB* = 60°, and prove that $AB = \frac{2}{3} OZ (2\sqrt{3} - 3)$.

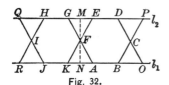

Fig. 32.

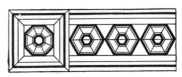

Fig. 32a. — Parquet Floor Border.

39. Figure 32a shows a series of regular hexagons inscribed between two parallel lines. Figure 32 is the diagram for the construction of Fig. 32a.

EXERCISES

1. In Fig. 32 find the number of degrees in each angle of △ *FNA*.

2. Prove that the line joining *F* and *C* is parallel to *RO*.

3. How must the points *E*, *D*, *P*, etc. and *A*, *B*, *O*, etc. be laid off from line *MN* in order that the hexagons may be constructed by drawing lines *DO*, *BP*, etc.?

4. If *MN* = 4, find the lengths of the sides of △ *FNA* and the length of *FC*. Find these lengths if *MN* = *a*.

$$Ans. \ \frac{a}{2}, \ \frac{a}{6}\sqrt{3}, \ \frac{a}{3}\sqrt{3}, \ \frac{2a}{3}\sqrt{3}.$$

5. If a rectangle is 16 inches wide, how long must it be if 16 hexagons are inscribed in it? *Ans.* 24.63 ft.

6. What must be the relation between the length and width of the rectangle in order that a whole number of regular hexagons may be inscribed in it?

7. What is the ratio of the sum of the areas of the hexagons to the area of the whole rectangle? *Ans.* ¾.

8. Does this ratio depend upon the shape or size of the rectangle?

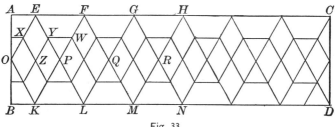

Fig. 33.

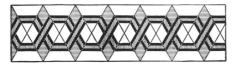

Fig. 33a. — Parquet Floor Border.

40. Figure 33 shows a series of rhombuses KPEO, LQFP, etc. inscribed between two parallel lines AC and BD. The angles of the rhombuses are 60° and 120°. The hexagons shown are formed by joining the middle points of the sides of the rhombuses.

EXERCISES

1. How must the points E, F, G, H, etc., and K, L, M, N, etc., be laid off on AC and BD in order that the rhombuses may be constructed by drawing the lines EL, FK, LG, etc.?

2. Prove that the hexagons shown are regular.

3. Prove that $XZYE$ is a rhombus.

4. If $AB = 8$, find the area of (1) $OKPE$, (2) one of the hexagons, (3) $XZYE$, and (4) $EYWF$. *Ans.* (1) $\frac{32}{3}\sqrt{3}$.

5. Find these areas if $AB = a$.

$$Ans.\ (1)\ \frac{a^2}{6}\sqrt{3};\ (2)\ \frac{a^2}{8}\sqrt{3};\ (3)\ \frac{a^2}{24}\sqrt{3};\ (4)\ \frac{a^2}{16}\sqrt{3}.$$

6. In Fig. 33a if the width of the border is 12 inches, find the total length of the dark bands in a border of 12 feet. Measure through the center between the bands.

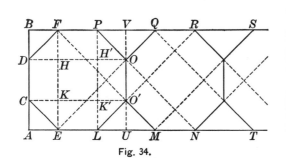

Fig. 34.

Fig. 34a. — Ceiling Pattern.

41. Figure 34 shows a series of regular octagons inscribed between two parallel lines.

EXERCISES

1. Show how to construct the series of octagons.

Suggestion. — Draw parallels BA, UV, etc., so as to construct a series of adjacent squares. In each square inscribe an octagon. See § 29.

2. If the octagon $CELO'$ etc. is regular, prove that CO' and DO are parallel to AU and BV, respectively.

3. Find the number of degrees in each angle of the △ CEK and DHF.

4. If one side CE of the regular octagon is 1 inch, find the lengths of CK and EK.   *Ans.* $\frac{1}{2}\sqrt{2}$.

5. Find lengths of these lines if $CE = a$.   *Ans.* $\frac{a}{2}\sqrt{2}$.

6. If $CE = 2$, find the area of the trapezoid $CEFD$, of the rectangle $ELPF$, and of the complete octagon $ELO'OPFDC$.

Ans. Area of octagon $8(1 + \sqrt{2})$.

7. Find these areas if $CE = a$.   *Ans.* $2\,a^2(\sqrt{2} + 1)$.

8. Prove that the points E, O, Q and also L, O', R are collinear.

9. If $EL = a$, find the distances AE, EL, LM, MN, NT, and so on, and lay off the points E, L, M, N, T and F, P, Q, R, S. How is LM related to AE? Construct the octagons. *Ans.* $\frac{a}{2}\sqrt{2}$, a, $a\sqrt{2}$, a, $a\sqrt{2}$.

10. What must be the relation between the length and width of a rectangle in order that a whole number of regular octagons may be inscribed in it?

11. If $AB = 4$, find AE and EL.

12. If $AB = a$, find these lengths.

$$Ans. \ AE = \frac{a}{2}(2 - \sqrt{2}), \ EL = a(\sqrt{2} - 1).$$

13. If $EL = b$, find AB. $\qquad$ *Ans.* $AB = b(\sqrt{2} + 1)$.

14. If $EL = b$, find the area of the octagon by subtracting the areas of the triangles AEC, LUO', OVP, and DFB from the area of the square $AUVB$. $\qquad$ *Ans.* Area of octagon $2\,b^2(1 + \sqrt{2})$.

15. If 12 regular octagons whose sides are each 4 inches are inscribed in a rectangle, find the dimensions of the rectangle.

Ans. 9.656; 115.8.

16. What is the ratio of the sum of the areas of the inscribed octagons to the area of the rectangle? $\qquad$ *Ans.* $2(\sqrt{2} - 1)$.

PART 2. DESIGNS BASED ON THE ISOSCELES RIGHT TRIANGLE

42. The designs in this section are all based on the unit shown in Fig. 35. It is interesting to note the number of times this unit is used in each design and the effect of the various positions in which the unit is placed.

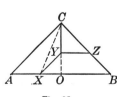

Fig. 35.

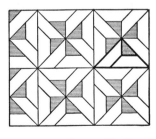

Fig. 35*a.* — Parquet Flooring.

43. In Fig. 35 ABC is an isosceles right triangle with a right angle at C. CO is perpendicular from C to AB.

EXERCISES

1. Show how to construct XY parallel to AC so that $CY = XY$.

Suggestion. — Bisect $\angle OCA$.

2. Draw YZ from Y parallel to AB and prove that $XY = CY$ $= YZ = BZ = AX$.

3. Show that the lines mentioned in Ex. 2 would have been equal if $\angle CBA$ or $\angle CAB$ had been bisected.

4. Prove that $BZYX$ and $AXYC$ are congruent trapezoids.

5. Prove that $OY = OX$, and hence show that $\dfrac{XY}{OY} = \sqrt{2}$.

6. Find the ratio $\dfrac{CZ}{BZ}$ and $\dfrac{CY}{OY}$.

7. Prove that XY extended is perpendicular to CB and that the altitude of $\triangle CYZ$ on CZ as base is equal to the altitude of trapezoid $AXYC$.

8. If $AC = 8$, find CY and the area of $AXYC$.

Ans. (1) $8(\sqrt{2} - 1)$; (2) $32(\sqrt{2} - 1)$.

Suggestion. — For the area of $AXYC$ show that $YO = 4(2 - \sqrt{2})$. Subtract the area of $\triangle XOY$ from the area of $\triangle AOC$.

9. If $AC = a$, find CY and the area of $AXYC$.

Ans. (1) $a(\sqrt{2} - 1)$; (2) $\dfrac{a^2}{2}(\sqrt{2} - 1)$.

10. If $AB = a$, make these same computations.

Ans. (1) $\dfrac{a}{2}(2 - \sqrt{2})$; (2) $\dfrac{a^2}{4}(\sqrt{2} - 1)$.

11. If $AB = a$, find the area of $\triangle CYZ$.   *Ans.* $\dfrac{a^2}{4}(3 - 2\sqrt{2})$.

12. If $AB = a$, show that the area of $AXYC$ is about $\dfrac{a^2}{10}$.

Suggestion. — $\dfrac{a^2(\sqrt{2} - 1)}{4}$ is $a^2(.101^+)$. From the figure, $AXYC$ is less than $\frac{1}{2} ABC$, and hence less than $\frac{1}{8} a^2$.

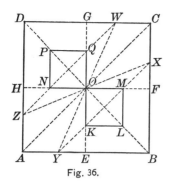

Fig. 36.

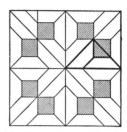

Fig. 36a. — Parquet Flooring.

44. In Fig. 36 ABCD is a square with the diameters EG and HF and the diagonals AC and DB. OX and OY bisect ∠s FOC and EOA, respectively. The line XY cuts OE and OF at K and M, respectively. KL and ML are drawn from K and M parallel to the sides AB and BC.

EXERCISES

1. Prove that XY is parallel to AC.

2. Prove that $OK = OM$ and that the two △ OKL and LMO form a square with the vertex L on the diagonal OB. Construct the figure.

3. If $AB = a$, find the area of the square OL. *Ans.* $\dfrac{a^2}{2}(3 - 2\sqrt{2})$.

4. What per cent of the design in Fig. 36a is made of dark wood?

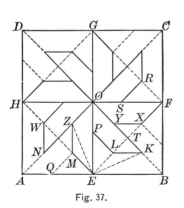

Fig. 37.

Fig. 37a. — Parquet Flooring.

45. In Fig. 37 ABCD is a square with the diagonal AC and BD and the diameters EG and HF.

EXERCISES

1. How many times is the unit shown in Fig. 35 used in this figure?

2. Show how to construct $EL = LP = LK = KB$. Show that this may be done by bisecting any one of three different angles.

3. Construct the entire figure.

4. Prove that points K, L, M, N; X, Y, Z, W; and Q, M, P, S, R are collinear.

5. If $AB = a$, find the area of $\triangle ELP$. *Ans.* $\frac{a^2}{8}(3 - 2\sqrt{2})$.

6. If $AB = a$, find the area of the wheel-shaped figure.

Ans. $a^2(\sqrt{2} - 1)$.

Suggestion. — Prove that $\triangle LKT = \triangle TX\dot{Y}$. Then the area of $OPLKYXFO$ is the area of $\triangle EFO$ less the area of $\triangle ELP$.

7. If $AB = a$, find the total area of the white triangles about one of the wheels in Fig. 37a. *Ans.* $a^2(3 - 2\sqrt{2})$.

8. Find the per cent of each kind of wood in one square in this design.

46. In Fig. 38 ABCD is a square with AC and BD the diagonals and EG and FH the diameters.

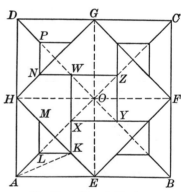

Fig. 38. — Parquet Flooring.

EXERCISES

1. Show how to obtain Fig. 38 from Fig. 37.

2. Show how to construct the lines in $\triangle AEH$ so that $LM = LK = LA$.

3. Construct the entire figure by repeating the construction suggested in Ex. 2, and then prove that the points *L, M, N, P* are collinear.

4. Give more than one method for constructing the lines shown in the square *EFGH*.

5. For each construction given in answer to Ex. 4 prove that $LK = KX = XO$.

6. Prove that *XYZW* is a square.

7. If $AB = a$, show that the area of *XYZW* is $\dfrac{a^2}{2}(3 - 2\sqrt{2})$.

PART 3. DESIGNS CONTAINING SQUARES WITHIN SQUARES

FIRST POSITION OF THE SECONDARY SQUARE

47. Figures 39 to 42 show the secondary square with its sides parallel to the diagonals of the given square. The most common form of the design is shown in Fig. 39. The square formed by joining the middle points of the sides of a given square is of universal occurrence. A frequent Roman design is made from Fig. 42. See § 52.

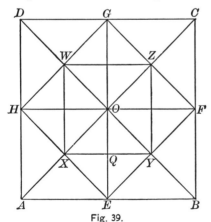

Fig. 39.

Fig. 39a. — Tile Pattern.

48. In Fig. 39 ABCD is a square with AC and BD its diagonals. E, F, G and H are the middle points of the sides and the points are joined as indicated.

EXERCISES

1. Prove that $EFGH$ is a square.

2. If the sides of square $EFGH$ cut AC and BD at points X, Y, Z, and W, prove that $XYZW$ is a square.

3. If $AB = 4$, find the area of the squares $EFGH$ and $XYZW$. Find the area of the trapezoid $ABYX$ and of the $\triangle XYO$. Add the areas of the trapezoid and of the triangle. The sum should be $\frac{1}{4}$ the area of square $ABCD$. Why?

4. If $AB = a$, find the area of (1) $EFGH$; (2) $XYZW$; (3) $ABYX$; (4) XYO.

$$Ans. \ (1) \ \frac{a^2}{2}; \ (2) \ \frac{a^2}{4}; \ (3) \ \frac{3\,a^2}{16}; \ (4) \ \frac{a^2}{16}.$$

5. How does the area of the $\triangle ABO$ compare with the area of the square $EFGH$? With the area of the square $XYZW$? Does this depend upon the length of AB?

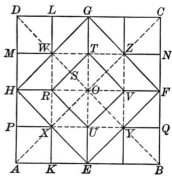

49. In Fig. 40 ABCD is a square with diameters and diagonals drawn. The points E, F, G, and H are joined. Various methods may be used to construct the remainder of the figure as indicated below.

Fig. 40. — Parquet Floor Design. Medieval Mosaic. Gayet, Vol. II, Parenzo, Pl. 32.

EXERCISES

1. Let L and K be the middle points of DG and AE, M and N the middle points of DH and CF, etc. Draw LK, MN, etc. Prove that (1) DO, GH, LK, and MN are concurrent, and (2) $MWLD$ is a square.

2. As an alternative method of construction, form squares in the $\triangle AEH$, EBF, etc. with AX, BY, etc. as diameters, and prove that the points K, X, W, and L are collinear.

3. As a third method, join W and X, X and Y, etc., and prove that (1) WX bisects HO and if extended bisects AE and GD, and (2) $MWLD$ is a square.

4. If the diameter GE cuts WZ and XY at T and U, respectively, and if the diameter HF cuts WX and ZY at R and V, respectively, prove that $HRTG$ is a trapezoid.

5. Prove that $RUVT$ is a square.

6. If $AB = 4$, find the length of WS.

7. If $AB = a$, find the length of WS.   *Ans.* $\dfrac{a}{8}\sqrt{2}$.

8. If $AB = a$, find the area of $HRTG$ by

(1) Finding the length of GH, RT, and WS, then using the formula for the area of a trapezoid;

(2) Finding the area of $OGDH$ and then subtracting the areas of HGD and OTR.   *Ans.* $\dfrac{3\,a^2}{32}$.

9. If $AB = a$, find the area of the square $MWLD$.   *Ans.* $\dfrac{a^2}{16}$.

10. If $AB = a$, find the area of the square $RUVT$.   *Ans.* $\dfrac{a^2}{8}$.

50. In Fig. 41 ABCD is a square, DB one diagonal, and E, F, G, H, the middle points of the sides.

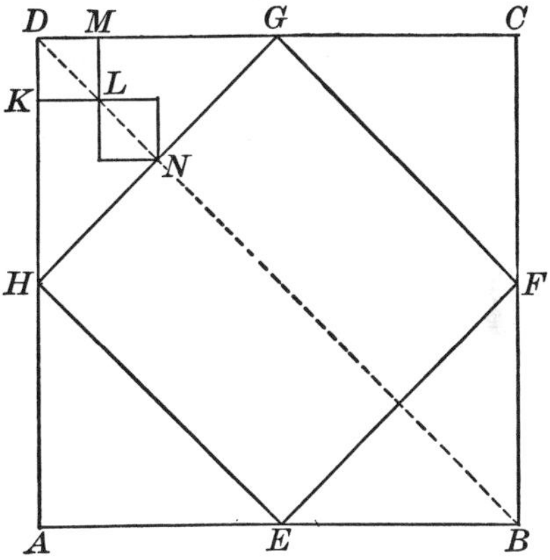

Fig. 41. — From Tiled Vestibule Design.

EXERCISES

1. Construct two equal squares between the points D and N so that the sum of their diagonals is DN.

2. If $AB = a$, find the area of the square $DKLM$.   *Ans.* $\dfrac{a^2}{64}$.

3. If the line $DL = s$, find the length of AB.   *Ans.* $4\,s\sqrt{2}$.

4. If n equal squares are constructed between D and N so that the sum of their diagonals is DN, find the area of each square if $AB = a$.

Ans. $\dfrac{a^2}{16\,n^2}$.

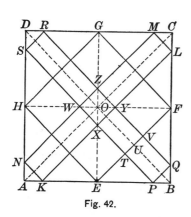

Fig. 42.

Fig. 42*a*. — Parquet Flooring.

51. In Fig. 42 ABCD is a square; E, F, G, H, the middle points of the sides; AK = BP = BQ = CL = CM, etc. ; and points are joined as shown.

EXERCISES

1. Prove that: (1) KL is parallel to AC; (2) $KLMN$ is a rectangle; (3) $XYZW$ is a square; (4) KL is parallel to EF; (5) points E, X, O, Z, and G are collinear.

Suggestion for (5). — EG is given a diameter of the square $ABCD$. Prove that $KX = PX$, and hence that X lies on EG.

2. What must be the relation of BP to AB in order that EF may be trisected by SP and QR?   *Ans.* $BP = \dfrac{AB}{6}$.

Suggestion. — Prove PQ parallel to EF.   Then prove

$$ET : TU :: EP : PB.$$

3. Draw a figure to illustrate the special case referred to in Ex. 2.

4. If EF is trisected by the lines PS and QR, prove that KL is divided into five equal parts by the lines EH, PS, QR, and FG.

5. If $AB = 12$ and $AK = 3$, find the length of KL and KN. Find the area of figure $AKLCMN$.

Ans. Area $= 63$.

6. If $AB = a$ and $AK = \dfrac{a}{n}$, find the lengths of (1) KN; (2) KL; (3) KP; and (4) PT. Find the area of (5) $AKLCMN$, and (6) of $KLMN$.

Ans. (1) $\dfrac{a}{n}\sqrt{2}$; (2) $\dfrac{a}{n}(n-1)\sqrt{2}$; (3) $\dfrac{a}{n}(n-2)$; (4) $\dfrac{a}{4\,n}(n-2)\sqrt{2}$;

(5) $\dfrac{a^2}{n^2}(2\,n-1)$; (6) $\dfrac{2\,a^2}{n^2}(n-1)$.

7. If $AB = a$ and if the area of the square $XYZW = \dfrac{a^2}{16}$, find AK in terms of a.

Ans. $\dfrac{a}{8}\sqrt{2}$.

8. If the area of $XYZW$ is $\dfrac{a^2}{n}$, find AK in terms of a and n.

Ans. $\dfrac{a}{\sqrt{2\,n}}$.

9. What must be the position of point K if the area of $KLMN$ is $\frac{1}{2}$ the area of the square $ABCD$?   *Ans.* $AK = \frac{1}{2}AB$.

Suggestion. — Solve for n the equation $\dfrac{2\,a^2}{n^2}(n-1) = \dfrac{a^2}{2}$.

10. What is the value of n if the area of $KLMN$ is $\dfrac{1}{k}$ the area of the square $ABCD$?   *Ans.* $k \pm \sqrt{k^2 - 2\,k}$.

Suggestion. — Solve for n the equation, $\dfrac{2\,a^2}{n^2}(n-1) = \dfrac{a^2}{k}$.

11. What must be the value of n if the area of $AKLCMN$ is $\dfrac{1}{k}$ of the area of the square $ABCD$?   *Ans.* $k \pm \sqrt{K^2 - K}$.

Fig. 43. — Roman Design.

52. Figure 43 is closely related to Fig. 42.

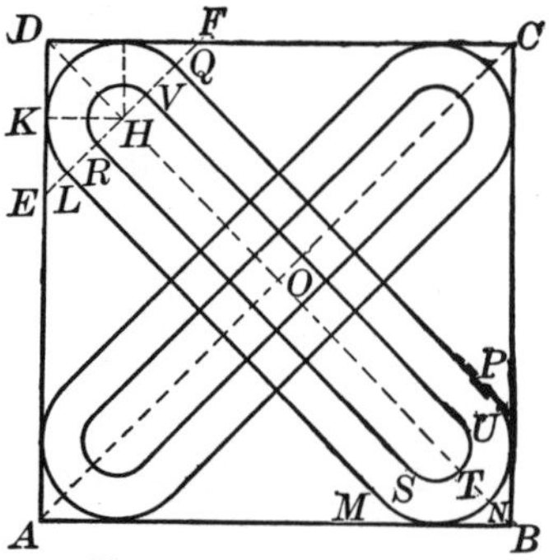

EXERCISES

1. Show how to construct Fig. 43.

Suggestion. — Make $DE = DF =$ etc.   Inscribe semicircles in $\triangle$ DEF, etc., so that their diameters shall be on the lines EF, etc., and be tangent to the sides of the square.

2. If $AB = a$ and $DE = \dfrac{a}{n}$, find (1) the radius of semicircle LKQ;

(2) the length of LM.   *Ans.* $\dfrac{a}{2\,n}, \dfrac{a}{n} \sqrt{2}(n-1)$.

3. If $AB = a$ and $DE = \dfrac{a}{n}$, find the area of the figure $LMNPQK$.

$$Ans. \ \frac{a^2}{n^2}\left(n\sqrt{2} - \sqrt{2} + \frac{\pi}{4}\right).$$

4. If $AB = 6$, $DE = 2$, and $HR = \frac{1}{4}$, find the area included between $LMNP$, etc. and $RSTU$, etc.   *Ans.* $6\sqrt{2} + \dfrac{15\,\pi}{16}$.

SECOND POSITION OF THE SECONDARY SQUARE

53. In Figs. 44–47 the sides of the secondary square are parallel to the sides of the given square.   Of the designs shown Fig. 47 is the most interesting, giving rise to an unusually large variety of patterns.

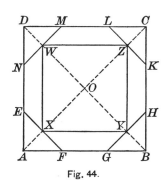

Fig. 44.

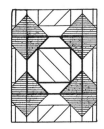

Fig. 44a. — Linoleum Pattern.

54. The design shown in square ABCD in Fig. 44 may be constructed by different methods, as indicated below.

EXERCISES

1. Construct $OX = OY = OZ = OW$. Join $X, Y, Z,$ and W. Draw EF, GH, etc., perpendicular to the diagonals at points X, Y, Z, etc. Prove that (1) $XYZW$ is a square; (2) XY is parallel to AB; (3) $AE = AF = GB$, etc.

2. Let W be any point on OD. From W draw WX parallel to DA, cutting AO at X; from X draw XY parallel to AB, etc. Prove that $OW = OX$.

3. As a third method of construction, make $AE = AF = GB$, etc. Let EF, GH, etc., intersect the diagonals at points X, Y, etc. Prove that (1) $XYZW$ is a square; (2) XY is parallel to AB; (3) EF is perpendicular to AO.

4. Prove that $XY = FB$.

5. If $AB = 16$ and $AF = 6$, find the areas of $XYZW$, $\triangle AFX$, and trapezoid $FGYX$.

6. If $AB = a$ and $AF = \dfrac{a}{n}$, find areas mentioned in Ex. 5.

$$Ans. \ (1) \ \frac{a^2}{n^2}(n-1)^2; \ (3) \ \frac{a^2}{4\,n^2}(2\,n-3).$$

7. Find the area of the trapezoid $FGYX$ (1) by using the formula for the area of the trapezoid, (2) by subtracting the areas of $\triangle AFX$, GBY, and XYO from the area of $\triangle ABO$.

8. If $AB = a$ and $AF = \dfrac{a}{n}$, find the area of trapezoid $FGYX$ when
(1) $n = 2$; (2) $\frac{3}{2} < n < 2$; and (3) $n < \frac{3}{2}$.

Suggestion. — Draw figures to illustrate each of these cases. When $n = 2$, the trapezoid becomes a triangle. For the other cases the difference between the areas of two triangles is found.

9. For what value of n in Ex. 6 will the area of trapezoid $FGYX$ be zero? Draw a figure to illustrate this case and prove it by geometry.

10. If the area of $\triangle AFE$ is equal to the area of $FGYX$, find the ratio of AF to AB.   *Ans.* $\frac{2}{3}$.

Suggestion. — Solve for n in the equation $\dfrac{a^2}{4\,n^2}(2\,n - 3) = \dfrac{a^2}{2\,n^2}$, where $AF = \dfrac{AB}{n}$.

11. If the area of the $\triangle AFE$ is equal to $\frac{1}{2}$ the area of the square $XYZW$, find the ratio of AF to AB.   *Ans.* $\frac{1}{2}$.

Suggestion. — Solve for n in the equation $\dfrac{a^2}{n^2}(n - 1)^2 = \dfrac{a^2}{n^2}$, where $AF = \dfrac{AB}{n}$.

12. Construct figures to illustrate the special cases mentioned in Exs. 10 and 11.

13. If the area of $\triangle AFE$ is equal to the area of $\triangle XOY$, find the ratio of AF to AB.   *Ans.* $\dfrac{1}{\sqrt{2} + 1}$.

14. Show that in the case mentioned in Ex. 13, $EF = XY$, and draw an illustrative figure.

Suggestion. — Divide AB in the ratio $1 : \sqrt{2}$; that is, in the ratio of a side of a square to its diagonal. See Fig. 44b, where $\dfrac{AF}{FB} = \dfrac{1}{\sqrt{2}}$ and $\therefore$ $\dfrac{AF}{AB} = \dfrac{1}{\sqrt{2} + 1}$.

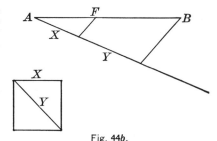

Fig. 44b.

15. In the case mentioned in Ex. 13 find the area of $XYZW$, if $AB = a$.

Ans. $2\,a^2(3 - 2\sqrt{2})$.

16. If $AB = a$, find the value of AF so that the area of $FGYX$ may be $\frac{1}{16}$ that of the square $ABCD$.   *Ans.* $\dfrac{AB}{6}$ or $\dfrac{AB}{2}$.

Suggestion. — Solve for n in the equation $\dfrac{a^2}{4\,n^2}(2\,n - 3) = \dfrac{a^2}{16}$. Then $AF = \dfrac{a}{n}$.

17. Construct a figure showing the meaning of each answer obtained in Ex. 16.

18. If $AB = a$, find AF so that the area of $FGYX$ is $\dfrac{1}{k}$ that of $ABCD$.   *Ans.* $\dfrac{a(k \pm \sqrt{k^2 - 12\,k})}{3\,k}$.

Suggestion. — Solve for n the equation $\dfrac{a^2(2\,n - 3)}{4\,n^2} = \dfrac{a^2}{k}$.

19. Show that the area of $FGYX$ cannot be more than $\frac{1}{12}$ that of of $ABCD$.

Suggestion. — The answer to Ex. 18 is imaginary if $k < 12$.

20. Show that the sum of the two values of AF obtained in Ex. 18 is $\frac{2}{3}$ of AB.

21. Draw a figure to illustrate the two values of AF obtained in Ex. 18.

Suggestion. — For the first value let F be any arbitrary point on AB so that $AF < \frac{2}{3} AB$. The other value of AF may be found from Ex. 20.

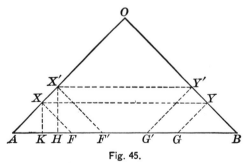

Fig. 45.

55. Some of the exercises given in the previous paragraph suggest interesting relations in isosceles triangles.

EXERCISES

1. If ABO is any isosceles triangle, and if $AF = BG$ and FX and GY are parallel to OB and OA, respectively, then XY is parallel to AB.

2. If $AB = a$ and $AF = h$, find the area of $FGYX$ when AOB is a right triangle.   *Ans.* $\dfrac{2\,ah - 3\,h^2}{4}$.

3. If ABO is a right triangle, find points F' and G' so that if $F'X'$ and $G'Y$ are drawn parallel to OB and OA, respectively, the area of $F'G'Y'X'$ shall be equal to that of $FGYX$.

Suggestion. — If $AF' = k$, the area of $F'G'X'Y' = \dfrac{2\,ak - 3\,k^2}{4}$.

Hence, $\dfrac{2\,ah - 3\,h^2}{4} = \dfrac{2\,ak - 3\,k^2}{4}$, and $h + k = \dfrac{2}{3}\,a$.

4. Study the case $h \gtrless \frac{1}{2} a$.

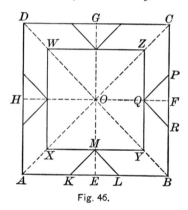

Fig. 46.

Fig. 46a. — Parquet Flooring.

56. In Fig. 46 ABCD is a square with diameters and diagonals drawn as shown. OX = OY = OZ = OW. Lines XY, YZ, etc. intersect the diameters at points M, Q, etc. From points M, Q, etc., lines MK, ML, QR, etc. are drawn parallel to the diagonals.

EXERCISES

1. Prove that (1) $MK = ML = QR$; (2) $KE = ME = RF$; (3) points K, M, Q, P are collinear; (4) $WXYZ$ is a square.

2. If $AB = 6$ and $KE = 1$, find the area of the square $XYZW$, of the trapezoid $ABYX$, and of the parallelogram $AKMX$.

3. If $AB = a$ and $KE = \dfrac{a}{n}$, find the area of the square $XYZW$, of the trapezoid $ABYX$, and of the parallelogram $AKMX$.

$$Ans. \ (1) \ \frac{a^2}{n^2}(n-2)^2; \ (2) \ \frac{a^2}{n^2}(n-1); \ (3) \ \frac{a^2}{2\,n^2}(n-2).$$

4. If $AB = a$ and $KE = \dfrac{a}{n}$, find n so that $KM = MX$.

$$Ans. \ 2 + 2\sqrt{2}.$$

Suggestion. —If $KM = KA$, then $\dfrac{KE}{KA} = \dfrac{1}{\sqrt{2}}$,

from which $\qquad\qquad \dfrac{KE}{AB} = \dfrac{1}{2 + 2\sqrt{2}}.$

The same result is obtained by solving $\dfrac{a\sqrt{2}}{n} = \dfrac{a}{2\,n}(n-2)$, for n.

5. Construct a figure to illustrate the case $KM = MX$.

Suggestion. — Divide AE in the ratio $1 : \sqrt{2}$. See § 54, Ex. 14.

6. If $AB = a$ and $KE = \dfrac{a}{n}$, find n so that $KM = XY$.

$$Ans. \ 2 + \sqrt{2}.$$

7. Construct a figure to illustrate the case $KM = XY$.

8. If the area of $ABCD$ is twice that of $XYZW$ and if $AB = 6$, find KE. $\qquad\qquad\qquad\qquad\qquad$ *Ans.* .88.

Suggestion. — Solve the equation $[6 - 2(KE)]^2 = 18$ for KE.

9. If $AB = a$, find KE so that the area of $ABCD$ shall be k times that of $XYZW$. $\qquad\qquad\qquad$ *Ans.* $\dfrac{a(\sqrt{k}-1)}{2\sqrt{k}}.$

10. If $AB = 6$, find KE so that the area of $AKMX$ shall be $\frac{1}{32}$ that of $ABCD$. $\qquad\qquad\qquad$ *Ans.* .4 or 2.5 nearly.

Suggestion. — Solve for KE in the equation $(3 - KE)KE = \frac{36}{32}$.

11. Construct a figure to show that both of the results obtained in Ex. 10 have a meaning.

12. If $AB = a$, find KE so that the area of $AKMX$ shall be $\dfrac{1}{k}$ that of $ABCD$. $\qquad\qquad\qquad$ *Ans.* $\dfrac{a(k \mp \sqrt{k^2 - 16\,k})}{4\,k}.$

13. Show that the area of the parallelogram $AKMX$ cannot be more than $\frac{1}{16}$ that of $ABCD$.

Suggestion. — If $AKMX$ were more than $\frac{1}{16}$ $ABCD$, the result obtained in Ex. 12 would be imaginary.

14. Show that the sum of the two values of KE obtained in Ex. 12 is $\frac{a}{2}$.

15. Given an isosceles right $\triangle AOE$ with right angle at E. Show how to construct KM and $K'M'$ parallel to AO, and XM and $X'M'$ parallel to AE so that the parallelograms $AKMX$ and $AK'M'X'$ shall be equal in area.

Suggestion. — From Ex. 12 and Ex. 14 it follows that $KE = AK'$ in Fig. 46b.

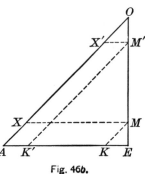

Fig. 46b.

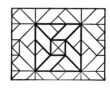

Fig. 47. — Parquet Floorings.

57. In Fig. 47 ABCD is a square with diagonals AC and BD. AE = BF = CG = DH. EY and GW are parallel to AC and FZ and HX parallel to BD.

EXERCISES

1. Prove that,

(1) $AEYX$ is a parallelogram.

(2) $AEYX \cong BFZY \cong CGWZ \cong DHXW$.

(3) $XYZW$ is a square.

2. If $AB = 10$ and $AE = 2$, find the areas of EBY, $AEYX$, and $XYZW$. *Ans.* 16, 8, 4.

3. If $AB = a$ and $AE = \dfrac{a}{n}$, find the areas mentioned in Ex. 2 and the perpendicular distance from Y to AB.

Ans. (1) $\dfrac{a^2}{4\,n^2}(n-1)^2$; (2) $\dfrac{a^2}{2\,n^2}(n-1)$; (3) $\dfrac{a^2}{n^2}$; (4) $\dfrac{a}{2\,n}(n-1)$.

4. If $AB = a$, find AE so that the area of $XYZW$ shall be $\dfrac{a^2}{n}$.

Ans. $\dfrac{a}{\sqrt{n}}$.

5. If $AB = a$, find CG so that $DG = 2\,WZ$. *Ans.* $\dfrac{a}{3}$.

6. If $AB = 6$, find AE so that the area of $AEYX$ shall be $\dfrac{a^2}{12}$.

Ans. 1.26 or 4.73 nearly.

Suggestion. — Solve for n the equation $\dfrac{a^2}{2\,n^2}(n-1) = \dfrac{a^2}{12}$.
Then $AE = \dfrac{6}{n}$.

7. Construct a figure to illustrate the meaning of the two results obtained in Ex. 6.

8. If $AB = a$, find AE so that the area of $AEYX$ shall be $\dfrac{a^2}{k}$.

Ans. $\dfrac{a\,(k \mp \sqrt{k^2 - 8\,k})}{2\,k}$.

9. Show that the parallelogram $AEYX$ cannot be more than $\frac{1}{8}$ the square $ABCD$.

10. Show that the sum of the two values of AE obtained in Ex. 6 is AB.

11. In connection with Ex. 5 construct the unit upon which the middle figure is based.

THIRD POSITION OF THE SECONDARY SQUARE

58. Designs containing secondary squares whose sides are not parallel to either the diagonals or the diameters of the given squares give still another variety. The Arabic unit shown in Fig. 48 gives rise to a great variety of patterns ancient and modern.[1]

[1] Day, pp. 45–47.

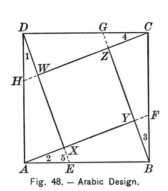

Fig. 48. — Arabic Design.

Fig. 48a. — Parquet Flooring.

59. The design shown in the square ABCD in Fig. 48 may be constructed by either method indicated below.

EXERCISES

1. $AE = BF = CG = DH$ and the points are joined as indicated. Prove that (1) $AFCH$ and $EBGD$ are congruent parallelograms; (2) $XYZW$ is a square.

Suggestion for (2). — Prove that (1) $\angle 2 + \angle 5 = 1$ rt. $\angle$; (2) $\angle WXY = 1$ rt. $\angle$; and (3) $XY = XW$.

2. As an alternative construction let $\angle 1 = \angle 2 = \angle 3 = \angle 4$. Prove that $XYZW$ is a square.

3. If $AB = a$ and $AE = \dfrac{a}{2}$, show that $AX = XY$.

Suggestion. — XE is parallel to YB in $\triangle ABY$.

4. If $AB = a$ and $AE = \dfrac{a}{2}$, prove that $YF = \dfrac{1}{2} YB = \dfrac{1}{2} AX = \dfrac{1}{5} AF$.

Suggestion. — $\triangle ABF$ is similar to $\triangle FBY$.

5. If $AB = a$ and $AE = \dfrac{a}{3}$, show that $AX = \dfrac{1}{2} XY$ and $YF = \dfrac{1}{9} AY$.

6. If $AB = a$ and $AE = \dfrac{a}{n}$, show that $AX = \dfrac{XY}{n-1}$ and $FY = \dfrac{AY}{n^2}$.

7. If $AB = a$ and $AE = \dfrac{a}{2}$, show that

(a) $AF = \dfrac{a}{2}\sqrt{5}.$

(b) $AX = \dfrac{a}{5}\sqrt{5} = \dfrac{2}{5} AF.$

(c) $AX = XY = BY.$

(d) $\overline{WX}^2 = \dfrac{a^2}{5}.$

8. If $AB = a$ and $AE = \dfrac{a}{3}$, show that

(a) $AF = \dfrac{a}{3}\sqrt{10}.$

(b) area of triangle $ABF = \dfrac{a^2}{6}.$

(c) $AX = \dfrac{a}{10}\sqrt{10}.$

(d) $EX = \dfrac{AX}{3} = \dfrac{AF}{10}.$

(e) $BY = \dfrac{3}{10} AF.$

(f) $XY = \dfrac{3}{5} AF = \dfrac{a}{5}\sqrt{10}.$

(g) Area of $XYZW = \tfrac{2}{5} a^2.$

9. If $AB = a$ and $AE = \dfrac{a}{n}$, show that

(a) $AF = \dfrac{a}{n}\sqrt{n^2 + 1}.$

(b) Area of triangle $ABF = \dfrac{a^2}{2n}.$

(c) $AX = \dfrac{a\sqrt{n^2 + 1}}{n^2 + 1}.$

(d) $EX = \dfrac{a\sqrt{n^2 + 1}}{n(n^2 + 1)}.$

Suggestion. — Since $\triangle AEX$ is similar to $\triangle ABF$, $EX : AX :: 1 : n.$

(e) $XY = \dfrac{a(n-1)}{(n^2 + 1)}\sqrt{n^2 + 1}.$

(f) $\overline{XY}^2 = \dfrac{a^2(n-1)^2}{n^2 + 1}.$

10. If $AB = 8$, find AY and BY so that the area of $XYZW$ shall be 32.   *Ans.* $b = 2(\sqrt{6} + \sqrt{2}), c = 2(\sqrt{6} - \sqrt{2}).$

Suggestion. — Use the equations $(b - c)^2 = 32$, and $b^2 + c^2 = 64$, where $b = AY$ and $c = BY.$

11. If $AB = a$, find AY and BY so that the area of $XYZW$ shall be $\dfrac{a^2}{n}.$ *Ans.* $\dfrac{a}{2}\left(\pm \sqrt{\dfrac{2n-1}{n}} + \sqrt{\dfrac{1}{n}} \right)$, $\dfrac{a}{2}\left(\pm \sqrt{\dfrac{2n-1}{n}} - \sqrt{\dfrac{1}{n}} \right).$

12. In the right triangle DWC, DC is the hypothenuse and $ABCD$ is the square on the hypothenuse. Draw the figure on pasteboard, cut it into pieces as indicated, and show that these can be rearranged so as to form the two squares on the sides DW and CW. Thus illustrate the Pythagorean theorem.[1]

[1] This proof is of Arabic origin. See Cajori, p. 123.

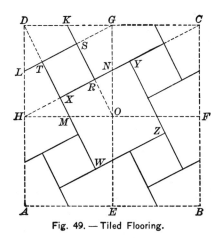

Fig. 49. — Tiled Flooring.

60. 1. Draw a diagram showing the construction for the tiled floor design in Fig. 49.

Suggestion. — In square $HOGD$, in Fig. 49, points K, L, M, and N are the middle points of the sides. The construction in adjacent squares is similar and similarly placed.

2. Prove that the points H, N, and C are collinear.

3. If, as an alternative construction, points H and C are joined, prove that HC bisects GO.

4. Find the ratio of a side of square $XRST$ to a side of the square $HOGD$. $Ans.\ \dfrac{\sqrt{5}}{5}.$

5. If four units identical with $HOGD$ are placed as shown in Fig. 49, prove that $XWZY$ is a square.

6. Prove that a side of square $XWZY$ is twice as long as a side of square $XRST$.

7. If the square $XWZY$ is $4\frac{1}{4}$ inches on a side, find the length of one side of square $HOGD$.   *Ans.* $4\frac{3}{4}$ nearly.

8. If a pavement contains 50 square yards, how many square yards are there of tiles like $RSTX$?   *Ans.* 10.

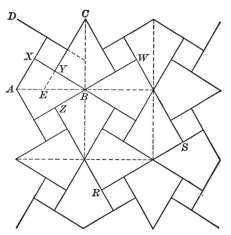

61. The unit on which the design in Fig. 50 is based is shown in the square ABCD. See Fig. 48. Notice the reversing of the position of the unit in adjacent squares.

Fig. 50. — Arabic Lattice.   Day, pp. 45 and 46.

EXERCISES

1. Prove that $ZRSW$ is a square.

2. If $AB = a$ and $AE = \dfrac{a}{n}$, show that $XY = \dfrac{BX}{n}$.

3. Find the ratio of XY to ZW under the conditions of **Ex. 2.**

$$Ans. \ \frac{1}{2\,n - 1}.$$

4. Find the value of n in Ex. 2 if $\overline{XY}^2 = \tfrac{1}{5}\overline{ZW}^2$.

Suggestion. — Solve for n in the equation $\dfrac{1}{(2\,n - 1)^2} = \dfrac{1}{5}$.

$$Ans. \ n = \tfrac{1}{2} \pm \tfrac{1}{2}\sqrt{5}.$$

5. Find the value of n in Ex. 2 if $\overline{XY}^2 = \dfrac{\overline{ZW}^2}{k}$.

$$Ans. \ n = \frac{1 \pm \sqrt{k}}{2}.$$

6. Show that CY, RZ, and AB are concurrent.

7. Find the area of the triangle CYB if $AB = a$.

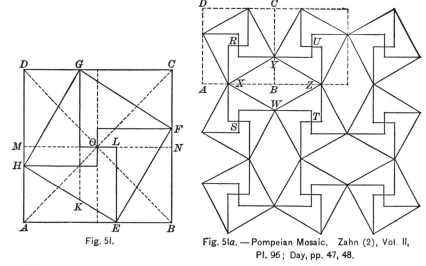

Fig. 5l.

Fig. 51a. — Pompeian Mosaic. Zahn (2), Vol. II,
Pl. 96; Day, pp. 47, 48.

62. In Fig. 51 ABCD is a square with its diagonals and diameters. AE=BF = CG = DH. GK and EL are perpendicular to MN and lines from H and F perpendicular to the second diameter.

EXERCISES

1. Prove that $EFGH$ is a square.

2. If $AE = \frac{2}{3} AB$, prove that the areas of the squares $ABCD$ and $EFGH$ are in the ratio 9 to 5.

3. If $AB = a$ and $BE = \dfrac{a}{n}$, prove that $\dfrac{a^2}{EF^2} = \dfrac{n^2}{n^2 - 2n + 2}$.

4. If $DG = \frac{1}{3} DC$, prove that $HK = KE$.

5. If $DG = \dfrac{DC}{n}$, prove that $HK = \dfrac{HE}{n-1}$.

6. If $AB = a$ and $EB = \dfrac{a}{n}$, find the area of $\triangle EBF$. *Ans.* $\dfrac{a^2}{2n^2}(n-1)$.

7. In Fig. 51a, if $AB = a$ and $AX = \dfrac{a}{n}$, find the area of the

diamond $XWZY$. *Ans.* $\dfrac{2a^2}{n^2}(n-1)$.

Suggestion. — Figure 51a is based on Fig. 51. The construction is shown by the dotted lines. Notice the reversing of the unit in adjacent squares.

8. Show that the area of the figure *RXSWTZUY* is equal to the area of the square *ABCD*.

9. Construct a design using the square with the inscribed swastika (see square *EFGH*, Fig. 51) as the unit. Reverse the position of the unit in adjacent squares.

Remark. — The design suggested in Ex. 9 is from the Alhambra.[1]

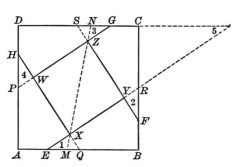

Fig. 52. — Arabic Design Unit. Bourgoin (2), Vol. 7, III, Pl. 14.

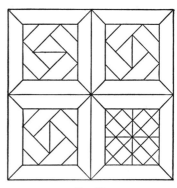

Fig. 52a.

63. In Fig. 52 *ABCD* is a square. AE = BF = CG = DH and AP = BQ = CR = DS. Points E and R, F and S, etc. are joined.

EXERCISES

1. Prove that *PG* and *ER* are parallel.

Suggestion. — $\triangle DPG \cong \triangle EBR$ and $\angle 1 = \angle 3 = \angle 5$.

2. Prove that:

(a) *EXQ* is a right angle.

Suggestion. — $\triangle EXQ$ is similar to $\triangle ERB$.

(b) *WX = XY.*

Suggestion. — $\triangle EXQ \cong \triangle FYR \cong \triangle PWH$ and $\triangle ERB \cong \triangle HAQ$

(c) Figures *EBFY* and *AEXH* are congruent.

(d) *XYZW* is a square.

[1] Calvert, (1), p. xxxviii. For the history of the swastika see Thomas Wilson, U. S. National Museum Report, 1894, p. 763.

3. As an alternative construction, make $AE = BF = CG = DH$ and $\angle 1 = \angle 2 = \angle 3 = \angle 4$.   Prove that (1) $AP = QB = CR = DS$ and (2) $XYZW$ is a square.

4. Find $\angle 1$ so that MN drawn through points X and Z is a diameter of square $ABCD$.   See Fig 52a.   *Ans.* 45°.

5. Show how the design shown in three of the squares in Fig. 52a may be derived from the one in the lower right-hand corner.

PART 4. DESIGNS CONTAINING STARS, ROSETTES, AND CROSSES WITHIN SQUARES

DESIGN CONTAINING THE RHOMBUS INSCRIBED IN THE SQUARE

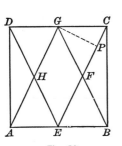

Fig. 53.

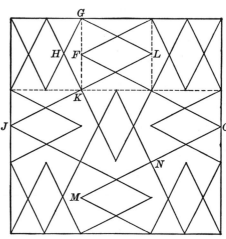

Fig. 53a. — Dutch Tile Design.

64. In Fig. 53, ABCD is a square. E and G are the middle points of AB and CD, respectively, and the points are joined as shown.

EXERCISES

1. Prove that $EFGH$ is a rhombus.

2. If $AB = 6$, find the length of EC.

3. Prove that $\triangle EBF$ and BCF are equal in area.   Find this area and the area of the rhombus $EFGH$ if $AB = 6$.

4. Find the areas mentioned in Ex. 3, if $AB = a$.

$$Ans. \ \frac{a^2}{8}, \ \frac{a^2}{4}.$$

5. Find the perpendicular distance between AG and CE, if $AB = a$.

$$Ans. \ \frac{a}{5}\sqrt{5}.$$

6. Construct the design shown in Fig. 53a. Notice the reversing of the position of the unit in adjacent squares.

7. Prove that in Fig. 53a points J, K, and L, also M, N, and O, are collinear.

8. In Fig. 53a show that the area of the rhombus FL is $\frac{2}{3}$ of the area of $KLGH$.

9. In Fig. 53a prove that $\angle LGH$ is a right angle.

DESIGN CONTAINING FOUR-POINTED STARS

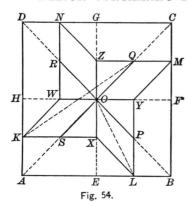

Fig. 54.

Fig. 54a. — Parquet Flooring.

65. In Fig. 54 ABCD is a square, with its diameters and diagonals. Points K, L, M, and N are the middle points of HA, EB, FC, and GD, respectively. KX, LY, MZ, and NW are drawn from the points K, L, M, and N perpendicular to the diameters, and the points joined as indicated.

EXERCISES

1. Prove that $LPOX$ and $KSOW$ are congruent parallelograms and that $\triangle PYO$ is equal in area to one half the parallelogram $LPOX$.

2. If $AB = a$, find the length of LO and KQ, and the area of $XLYMZ$ etc.

$$Ans. \ \frac{a^2}{4}\sqrt{5}, \ \frac{a^2}{4}\sqrt{13}, \ \frac{3\,a^2}{8}.$$

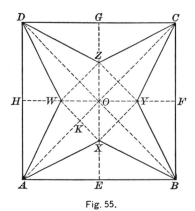

Fig. 55.

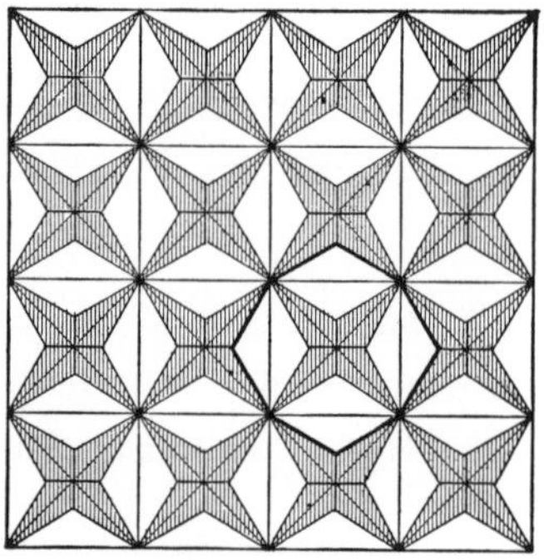

Fig. 55a. — Parquet Flooring.   Arabic
and Medieval.   Bourgoin (2), Vol. 7,
III, Pl. 6 ; Wyatt, Pl. 12.

66. In Fig. 55 ABCD is a square with diameters and diagonals. EX = FY
= GZ = HW, and the points are joined as indicated.

EXERCISES

1. Prove that (1) $XYZW$ is a square, (2) ABX and BCY are
congruent isosceles triangles, (3) AX is parallel to CZ.

2. If $EX = OX$, prove that $\triangle AXO$ and AEX are equal in area.

Suggestion. — Triangles having equal bases and equal altitudes are
equal in area.

3. If $EX = OX$, prove that the area of the star $AXBY$ etc. is $\frac{1}{2}$
that of the square $ABCD$.

4. If $AB = a$ and $EX = \frac{a}{n}$, prove that $\frac{OX}{OE} = \frac{n-2}{n}$ and $\frac{OX}{EX} = \frac{n-2}{2}$.

5. If $AB = a$ and $EX = \frac{a}{n}$, what is the relation of the area of
$\triangle AXO$ to the area of $\triangle AEO$, and the area of the star $AXBYC$ etc.
to the area of the square $ABCD$?   *Ans.* $\frac{n-2}{n}$.

6. What is the relation of EX to OE if the area of the star is $\frac{1}{4}$
of the area of the square?

7. What is the relation of EX to OE if the area of the star is $\dfrac{1}{k}$ of the area of the square $ABCD$? *Ans.* $\dfrac{k-1}{k}.$

8. If $AB = a$ and $EX = XO$, find the area of the $\triangle AXW$ and of the square $XYZW$. *Ans.* $\dfrac{3\,a^2}{32}$ and $\dfrac{a^2}{8}.$

9. If $AB = a$ and if $EX = \dfrac{a}{n}$, find lengths of OX, XW, and AK.

Ans. (1) $\dfrac{a(n-2)}{2\,n}$ (2) $\dfrac{a\sqrt{2}}{2\,n}(n-2)\,;$ (3) $\dfrac{a\sqrt{2}(n+2)}{4\,n}.$

10. Find the areas of $\triangle AXW$, square $XYZW$, and of the star if $AB = a$ and $EX = \dfrac{a}{n}.$

Ans. (1) $\dfrac{a^2}{8\,n^2}(n^2-4)\,;$ (2) $\dfrac{a^2}{2\,n^2}(n-2)^2\,;$ (3) $\dfrac{a^2(n-2)}{n}.$

11. If $AB = a$, what must be the length of EX in order that the area of $XYZW$ shall be $\frac{1}{12}$ of the area of the whole square?

Ans. $EX = \dfrac{5\,a}{12 \pm 2\sqrt{6}}.$

Suggestion. — Solve the equation $\dfrac{a^2(n-2)^2}{2\,n^2} = \dfrac{a^2}{12}.$

12. Have both results obtained in the answer to Ex. 11 a meaning in the figure?

Suggestion. — The values of EX and EZ are obtained.

13. What must be the value of EX if the square $XYZW$ is $\dfrac{1}{k}$ of the whole square? *Ans.* $\dfrac{(k-2)a}{2(k \pm \sqrt{2\,k})}.$

14. If $AB = a$, find EX so that the sum of the areas of the $\triangle AXW$, XBY, YCZ, and ZDW shall be $\frac{1}{10}$ of the area of the whole square.

Ans. $EX = \dfrac{a\sqrt{5}}{5}.$

15. Construct a figure to illustrate the case mentioned in **Ex. 14.**

Suggestion. — Since $\dfrac{\sqrt{5}}{a} = \dfrac{1}{EX}$, EX is the fourth proportional to $\sqrt{5}$, a, and 1. If a right triangle be constructed with the legs in the ratio $1:2$, the shorter leg and hypothenuse are in the ratio $1:\sqrt{5}$.

16. If $AB = a$, find EX so that the sum of the areas of the △ AXW, XBY, etc., shall be $\dfrac{1}{k}$ of the area of the square.

$$Ans. \ \frac{a}{2}\sqrt{\frac{k-2}{k}}.$$

17. What must be the length of EX if the area of the star is $\frac{1}{4}$ that of $ABCD$? Construct an illustrative figure.

18. What must be the length of EX if the area of the star is $\dfrac{1}{k}$ of that of $ABCD$? $Ans. \ \dfrac{a(k-1)}{2\,k}.$

19. If AX, BX, BY, etc. bisect the △ OAE, OBE, OEF, etc., prove that OE, AX, and BX are concurrent.

Suggestion. — The bisectors of the angles of a triangle are concurrent.

20. If AX, BX, BY, etc. bisect the △ OAE, OBE, etc., prove that $OX = OY = OZ = OW$.

21. If AX, BX, BY, etc. bisect △OAE, etc., find the lengths of EX, OX, and AX, when $AB = a$.

$$Ans. \ (1) \ \frac{a}{2}(\sqrt{2}-1); \ (2) \ \frac{a}{2}(2-\sqrt{2}); \ (3) \ \frac{a}{2}\sqrt{4-2\sqrt{2}}.$$

Suggestion. — The bisector of an angle of a triangle divides the opposite sides into parts proportional to the other two sides.

22. Also find the area of the star $AXBYC$ etc., and of △ AXB.

$$Ans. \ (1) \ a^2(2-\sqrt{2}); \ (2) \ \frac{a^2}{4}(\sqrt{2}-1).$$

23. Construct a diagram for a pattern like Fig. 55a in which the octagons formed shall be regular.

Suggestion. — Show that in Fig. 55 if AX, BX, BY, etc. bisect the △ OAE, OBE, etc., the octagons shown in Fig. 55a will be regular.

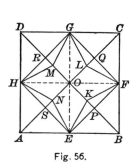

Fig. 56.

Fig. 56a. — Parquet Flooring.

67. In Fig. 56 ABCD is a square with diameters and diagonals. $OK = OL = OM = ON$ and the points are joined as shown.

EXERCISES

1. Prove that (1) $EK = KF = FL$, etc.; (2) $\angle MHN = \angle NEK = \angle KFL$, etc.; (3) $ONEK$ and $OKFL$ are congruent.

2. If $AB = 12$ and $OK = 3$, find the area of the star $EKFLG$ etc.

3. If $OK = \dfrac{OB}{n}$, compare the areas of $\triangle EKO$ and EBO, and prove that the area of the star is $\dfrac{1}{n}$ of that of the square $ABCD$.

4. If $AB = a$ and $OK = \dfrac{OB}{n}$, find the area of $ONEK$. *Ans.* $\dfrac{a^2}{4n}$.

5. Prove that $\dfrac{KN}{AB} = \dfrac{OK}{OB}$.

6. If $AB = 6$ and $KP = 1$, find the area of the star.

$Ans.\ 6\,(3 - \sqrt{2}).$

7. If $AB = a$ and $KP = \dfrac{OP}{n}$, find the area of $\triangle EFK$ and of the star. $Ans.$ (1) $\dfrac{a^2}{8n}$; (2) $\dfrac{a^2}{2n}(n - 1).$

8. Compare $\triangle EKF$ and EBF and show that the formula for the area of $\triangle EFK$ accords with the theorem: "The areas of triangles having equal bases are to each other as their altitudes."

9. If the area of the star is $\dfrac{1}{k}$ of the whole square, find PK.

$$Ans.\ PK = \frac{OP(k-2)}{k}.$$

Suggestion. — Use equation $a^2\left(\dfrac{n-1}{2\,n}\right) = \dfrac{a^2}{k}.$

10. If the sum of the areas of the △ EFK, FGL, etc. is $\dfrac{1}{k}$ of the whole square, find PK. $Ans.\ KP = \dfrac{2\,PO}{k}.$

11. Construct a diagram illustrating the conditions given in Ex. 10 if $k = 6$ and show that the area of the star is $\dfrac{a^2}{3}$.

68. Figure 57 is based upon a special case of Fig. 56. The unit of which it is composed is shown in square ABCD, where EK, KF, FL, etc. bisect the ∠ OEP, OFP, OFQ, etc., respectively.

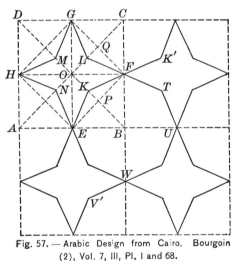

Fig. 57. — Arabic Design from Cairo. Bourgoin (2), Vol. 7, III, Pl. I and 68.

EXERCISES

1. Prove that EK, KF, and OB are concurrent.

Suggestion. — The bisectors of the angles of a triangle are concurrent.

2. Prove that $OK = OL = OM = ON$.

3. If $AB = a$, find the length of OK, the area of $EKFLG$ etc. and of $EKFB$.

$$Ans.\ (1)\ \frac{a}{2}(\sqrt{2}-1);\ (2)\ \frac{a^2}{2}(2-\sqrt{2});\ (3)\ \frac{a^2}{8}\sqrt{2}.$$

4. Construct the pattern shown in the figure.

5. Prove that (1) the points KFK' are collinear; (2) KFK' and $V'W$ are parallel; (3) $EWUTF$ etc. is a regular octagon.

6. Compare the area of $EWUTF$ etc. with the area of $EKFLG$ etc. *Ans.* Area $= (1 + \sqrt{2})$ times the area of star.

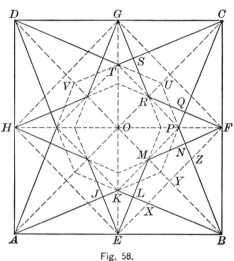

69. In Fig. 58 AK, KB, BP, etc. bisect the $\angle OAE$, EBO, OBF, etc., and EM, MF, FR, etc. bisect the $\angle OEF$, OFE, OFG, etc. These bisectors are extended until they intersect.

Fig. 58.

EXERCISES

1. Prove that

(a) $KL = NP = PQ$, etc.; (b) $EL = NF = FQ$, etc.; (c) $BL = BN = CQ$, etc.; (d) $LM = MN = QR$, etc.; (e) BK is the perpendicular bisector of EM; (f) $EK = EX$; (g) $OM = EX$ and $EM = XB$.

2. Prove that the figures $ELKJ$ and $NFQP$ are congruent; also figures $BNML$ and $CSRQ$. Prove that $\triangle ELK$ and EBL are similar.

3. If $AB = a$, show that

(1) $EK = \dfrac{a}{2}(\sqrt{2} - 1) = OM.$

(2) $OK = \dfrac{a}{2}(2 - \sqrt{2}).$

(3) $KB = \dfrac{a}{2}\sqrt{4 - 2\sqrt{2}}.$

(4) $MY = \dfrac{a}{4}(2 - \sqrt{2}).$

(5) $EM = \dfrac{a}{2}\sqrt{2 - \sqrt{2}}.$

(6) $KL = \dfrac{a}{4}\sqrt{10 - 7\sqrt{2}}.$

(7) $LB = \dfrac{a}{4}\sqrt{2 + \sqrt{2}}.$

4. Verify the values of *KL* and *LB* by showing that
$$(KL + LB)^2 = \overline{KB}^2.$$

5. Show also that the values given in Ex. 3 are in accord with the following:

(1) $\overline{EK}^2 = \overline{KL}^2 + \overline{LE}^2.$ (3) $\overline{EB}^2 = \overline{EL}^2 + \overline{LB}^2.$

(2) $\overline{EK}^2 = (KL)(KB).$ (4) $\overline{EB}^2 = (LB)(KB).$

(5) $\overline{EL}^2 = (KL)(LB).$

6. Extend sides *AK*, *KB*, *BP*, etc. until they intersect. Prove that the lines *DT*, *BP*, and *OC* are concurrent. Prove that a regular octagon is formed.

7. Extend the sides *EM*, *MF*, *FR*, etc. until they intersect. Prove that the lines *GR*, *EM*, and *OF* are concurrent. Prove that a regular octagon is formed whose sides are parallel to sides of octagon *VTUP* etc.

70. In Fig. 59 E, F, G, and H are the middle points of the sides of the square ABCD, and the points are joined as shown.

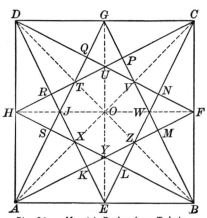

Fig. 59. — Moorish Design from Toledo.

EXERCISES

1. Prove that the points *E*, *Y*, *O*, *U*, and *G*, and also *B*, *Z*, *O*, *T*, and *D* are collinear.

Suggestion. — Prove ∠ *YAB* = ∠ *YBA* ; *AY* = *BY*; *GE* the perpendicular bisector of *AB*. Therefore *Y* lies on *EG*.

2. Prove that

(*a*) $EL = MF = FN$, etc.　　(*b*) $BL = BM = NC$, etc.

(*c*) $YZ = ZW = WV$, etc.　　(*d*) $LZ = ZM = VN$, etc.

(*e*) $KY = YL = WM$, etc.

3. Prove that $KMPR$ and $LNQS$ are equal squares, each equal to $\frac{1}{5}$ of the square $ABCD$. See §59, Ex. 7.

4. Show that the following figures are congruent:

(*a*) $BLZM$, $CNVP$, etc.; (*b*) $EKYL$, $FMWN$, etc.; (*c*) YLZ, ZWM, WVN, etc.

5. Show that the following triangles are similar:

(*a*) EBC, ELB, ELY, and EUN.　(*e*) BOY and BHD.

(*b*) BLZ, BST, and BXO.　(*f*) BEZ and ATB.

(*c*) EYZ and EUC.　(*g*) BYZ and BHT.

(*d*) YLZ, YMB, and BHP.　(*h*) BYF and BHC.

6. Show that if $AB = a$,

(*a*) $EC = \frac{a}{2}\sqrt{5}.$　(*c*) $LB = \frac{a}{5}\sqrt{5}.$　(*e*) $OZ = \frac{a}{6}\sqrt{2}.$　(*g*) $ZL = \frac{a}{15}\sqrt{5}.$

(*b*) $EL = \frac{a}{10}\sqrt{5}.$　(*d*) $LY = \frac{a}{20}\sqrt{5}.$　(*f*) $ZB = \frac{a}{3}\sqrt{2}.$　(*h*) $ZY = \frac{a}{12}\sqrt{5}.$

7. Is $XYZWVU$ etc. a regular octagon?

DESIGNS CONTAINING FOUR-PART ROSETTES

71. On the sides of the square shown in Fig. 60, $AG = HB = BK = CL$, etc. The points are joined as indicated.

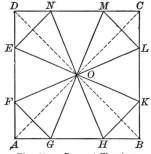

Fig. 60. — Parquet Flooring.

1. Prove that △ FGO, HKO, etc., also △ GHO, KLO, etc. are congruent and isosceles.

2. Show that the ratio of the area of $\triangle\, OGH$ to that of $\triangle\, OAB$ is as GH to AB.

3. Construct a figure in which the sum of the areas of △ GHO, KLO, etc. shall be ¼ the area of the square $ABCD$.

4. If $AB = a$, and $AG = \dfrac{a}{n}$, find the sum of the areas of the △ GHO, KLO, etc. $Ans.\ \dfrac{a^2}{n}\,(n-2).$

5. Find n in **Ex. 4** so that the sum of the areas of these triangles is $\dfrac{1}{k}$ of the square $ABCD$. $Ans.\ n = \dfrac{2\,k}{k-1}.$

6. Substitute 4 for k in the answer to **Ex. 5**, and show that the result agrees with the answer to **Ex. 3**.

7. If $AB = 8$ and $AG = 2$, find the sum of the areas of the △ FGO, HKO, etc. $Ans.\ 24.$

8. If $AB = 8$ and $AG = \dfrac{8}{n}$, find the sum of the areas of the △ FGO, HKO, etc. $Ans.\ \dfrac{128}{n^2}\,(n-1).$

9. If $AB = 8$, find AG so that the sum of these areas shall be ⅓ of the area of the square $ABCD$.

Suggestion. — Solve for n in the equation $\dfrac{128(n-1)}{n^2} = \dfrac{64}{3}.$

$Ans.\ \dfrac{8}{3 \pm \sqrt{3}}.$

10. Show that the two results obtained in **Ex. 9** are the lengths of AG and AH.

11. If $AB = a$ and $AG = \dfrac{a}{n}$, find the sum of the areas of these triangles. $Ans.\ \dfrac{2\,a^2}{n^2}(n-1).$

12. If $AB = a$, find AG so that the sum of the areas of these triangles shall be $\dfrac{1}{k}$ of the whole square.

$Ans.\ AG = \dfrac{a}{k \pm \sqrt{k^2 - 2\,k}}.$

13. Study the case $k = 2$.

14. If $AB = a$ and $AG = \dfrac{a}{n}$, find n so that $\triangle\, HKO$ and GHO are equal in area.

Suggestion. — Solve for n in the equation $\dfrac{n-2}{4\,n} = \dfrac{n-1}{2\,n^2}$.

$$Ans.\ n = 2 \pm \sqrt{2}.$$

15. Show that if $\triangle\, HKO$ and GHO are equal, the octagon $FGHKL$ etc. is regular.

16. If $AB = a$ and $AG = \dfrac{a}{n}$, find n so that $\triangle\, GHO$ and HBK are equal in area.

Suggestion. — Solve for n in the equation $\dfrac{n-2}{4\,n} = \dfrac{1}{2\,n^2}$.

$$Ans.\ n = 1 \pm \sqrt{3}.$$

17. Construct a figure to illustrate this case.

Suggestion. — Divide AB in the ratio $1 : \sqrt{3}$; that is, in the ratio of the half side to the altitude of the equilateral triangle.

18. If $AB = a$ and $AG = \dfrac{a}{n}$, find n so that $\triangle\, HKO$ and HBK shall be equal in area. $Ans.\ n = 2.$

19. If $AB = a$ and $AG = \dfrac{a}{n}$, find n so that the area of $\triangle\, GHO$ shall be twice that of $\triangle\, HBK$. $Ans.\ n = 1 \pm \sqrt{5}.$

20. Construct a figure to illustrate the case mentioned in Ex. 19.

Suggestion. — Construct a right $\triangle$ with the legs in the ratio $1 : 2$. Then the shorter leg is to the hypothenuse as $1 : \sqrt{5}$. Divide AB in the ratio $1 : \sqrt{5}$.

72. In Fig. 61 EFGH is the square formed by joining the middle points of the sides of the square ABCD. HK = ME = EN = FQ, etc., and the points are joined as indicated.

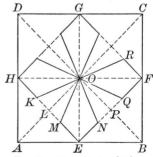

Fig. 61. — From Design for Iron Balcony Rail.

1. Prove that (1) $\triangle KMO \cong \triangle NQO$, etc., and (2) $ENOM \cong OQFR$, etc.

2. If $KL = \dfrac{HL}{n}$, find the ratio of the area of $\triangle KMO$ to that of $\triangle HEO$, and the ratio of the area of $HKOMENO$ etc. to that of square $EFGH$, and to that of square $ABCD$.

$Ans.$ $\dfrac{n-1}{2n}$ times square $ABCD$.

3. If $AB = 6$ and $KL = \dfrac{HE}{6}$, find the area of $HKOMENO$ etc.

$Ans.$ 12.

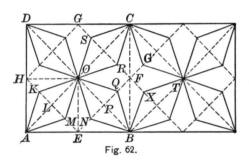

Fig. 62.

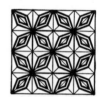

Fig. 62a. — Parquet Flooring. Alhambra Design. Jones (2), Pl. XLIII.

73. The design shown in Fig. 62 is derived from that shown in Fig. 61.

1. Prove that (1) $AMOK$ and $NBQO$ are congruent parallelograms, and (2) $OMEN \cong FROQ$.

2. Find the ratio of the area of $AMOK$ to that of square $AEOH$.

$Ans.$ $\dfrac{KM}{HE}$.

3. Construct a figure in which the sum of the areas of $AMOK$, $NBQO$, etc., shall be $\frac{1}{3}$ that of the square $ABCD$.

4. If $AB = a$ and $LM = \dfrac{HE}{4}$, find the area of $AMOK$. $Ans.$ $\dfrac{a^2}{8}$.

5. If $AB = a$ and $LM = \dfrac{HE}{n}$, find the sum of the areas of $AMOK$, $BQON$, etc., in square $ABCD$. $Ans.$ $\dfrac{2a^2}{n}$.

6. If $AB = a$ and $LM = \dfrac{HE}{n}$, find the area of $BXTGCR$ etc.

$$Ans. \ \frac{a^2(n-2)}{2\,n}.$$

7. For what value of n will the area of $BXTGCR$ etc. be zero?

8. If the sum of the areas of $AMOK$, $BQON$, etc. be $\dfrac{1}{k}$ of the given square, find LM. $\qquad Ans. \ \dfrac{HE}{2\,k}.$

9. If $OE = EK$, prove that $ABQOKA$ is one half of a regular octagon.

10. If $AB = a$ and $OE = EK$, find (1) the length of KL, (2) the area of $AMOK$, and (3) the area of $BQORCG$ etc.

$$Ans. \ \frac{a}{4}(2 - \sqrt{2}), \ \frac{a^2}{4}(\sqrt{2} - 1), \ \frac{a^2}{2}(2 - \sqrt{2}).$$

DESIGNS CONTAINING THE MALTESE AND RELATED CROSSES

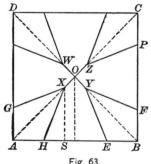

Fig. 63.

Fig. 63a. — Parquet Flooring.

74. On the sides of the square ABCD, shown in Fig. 63, GA = AH = BE = BF, etc. OX = OY = OZ = OW, and the points are joined as indicated.

EXERCISES

1. Prove that (1) $AHXG \cong EBFY$, (2) $HEYOX \cong FPZOY$, and (3) GX and ZP are parallel.

2. If $AB = a$, $AH = \dfrac{a}{4}$, and $XO = \dfrac{AO}{4}$, find the area of $AHXG$ and of $GXHEYF$ etc.   *Ans.* $\dfrac{3\,a^2}{32}$, $\dfrac{5\,a^2}{8}$.

3. If $AB = a$ and $OX = \dfrac{OA}{n}$, show that $XS = \dfrac{a(n-1)}{2\,n}$.

4. If $AB = a$, $AH = \dfrac{a}{n}$, and $OX = \dfrac{AO}{n}$, find the area of the figure $GXHEYF$ etc.   *Ans.* $\dfrac{a^2}{n^2}(n^2 - 2\,n + 2)$.

5. If $AB = a$, $AH = \dfrac{a}{n}$, and $XO = \dfrac{AO}{m}$, find the area of the figure $GXHEYF$ etc.   *Ans.* $\dfrac{a^2}{mn}(mn - 2\,m + 2)$.

6. Show how to obtain the answer to Ex. 4 from that to Ex. 5. Can the answer to Ex. 5 be obtained from that to Ex. 4?

7. If the cross $GXHEYF$ etc. occupies $\frac{1}{2}$ the area of the square $ABCD$, and if $XO = \dfrac{AO}{4}$, find AH, when $AB = a$.   *Ans.* $\dfrac{a}{3}$.

8. If $AB = a$, $OX = \dfrac{AO}{m}$, and $AH = \dfrac{a}{n}$, find the relation between m and n in order that the cross may occupy $\frac{1}{3}$ of the square.

$$Ans.\ n = \frac{3\,m - 3}{m}.$$

Suggestion.— Solve for n in the equation $\dfrac{a^2(mn - 2\,m + 2)}{mn} = \dfrac{a^2}{3}$.

9. In Ex. 8 if XO is zero, what is the value of AH? Construct a figure to illustrate this case.   *Ans.* $\dfrac{a}{3}$.

10. In Ex. 8 if $AH = \dfrac{a}{2}$, what is the value of XO? Construct an illustrative figure.   *Ans.* $\dfrac{OA}{3}$.

11. What relation must exist between m and n in order that the area of the cross shall be $\dfrac{1}{k}$ of the square?   *Ans.* $n = \dfrac{2\,(km-1)}{m\,(k-1)}$.

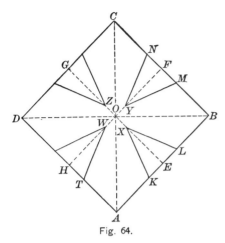

Fig. 64.

Fig. 64a. — Parquet Flooring.

75. On the sides of the square ABCD, shown in Fig. 64, **AK = BL = BM,** etc. **OW = OX = OY = OZ**, and the points are joined as shown.

EXERCISES

1. Prove that (1) $\triangle KLX \cong \triangle MYN$, (2) $AKXOWT \cong LBMYOXL$.

2. If $AB = a$, $KE = \dfrac{a}{6}$, and $OX = \dfrac{OE}{4}$, find the area of the figure $AKXLBMY$ etc.

$$Ans. \ \frac{3\,a^2}{4}.$$

3. If $AB = a$, $KE = \dfrac{a}{n}$, and $OX = \dfrac{OE}{m}$, find the area of the figure $AKXLBMY$ etc.

$$Ans. \ \frac{a^2}{mn}\,(mn - 2\,m + 2).$$

4. Notice that the answer given to Ex. 3 is the same as that to § 74, Ex. 5. Show by geometry that the crosses in Figs. 63 and 64 have the same area.

Suggestion. — See Fig. 64b and use the theorem that triangles having equal bases and equal altitudes are equal in area.

5. If $AB = a$ and $OX = \dfrac{OE}{4}$, find EK so that the area of $AKXLBM$ etc. shall be $\frac{1}{2}$ the area of the square $ABCD$.

$$Ans. \ EK = \frac{a}{3}.$$

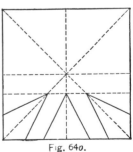

Fig. 64b.

6. If $AB = a$, $EK = \dfrac{a}{n}$, and $OX = \dfrac{OE}{m}$, find the relation between n and m so that $AKXLBM$ etc. shall be $\dfrac{1}{k}$ of the whole square.

$$Ans. \quad n = \frac{2\,k(m-1)}{m(k-1)}.$$

7. Show the relation between Fig. 64 and Fig. 60.

Ans. If XO is zero, Fig. 64 becomes Fig. 60.

8. Show the relation between Fig. 64 and Fig. 55.

Ans. If $KE = \dfrac{AB}{2}$, Fig. 64 becomes Fig. 55.

9. Join points X, Y, Z, and W; also T and K, L and M, etc. Prove that WX is parallel to TK and find the area of $TKXW$ if $AB = a$, $AK = \dfrac{a}{n}$, and $OX = \dfrac{OE}{m}$.

$$Ans. \quad \frac{a^2}{8\,m^2 n^2}\,(2\,mn - 2\,m - n)(2\,m + n).$$

Remark. — The figure suggested in Ex. 9 above is found in a mosaic in St. Mark's Cathedral, Venice.[1]

[1] See Dictionary of Architecture, plate opposite page 65 in the article on Pavements.

CHAPTER III

MISCELLANEOUS INDUSTRIES

PART 1. DESIGNS BASED ON OCTAGONS WITHIN SQUARES

76. Of the industries referred to in this chapter, ornamental iron and painted glass date back to the middle ages. The origin of mosaics and baked clay is lost in antiquity. On the other hand many of these industries are strictly modern.

77. Occurrence. — While numerous tile and parquet floor designs occur in this and the following Part, there are also given a number of designs from other industries, especially of mosaic floorings and linoleums. These designs are less simple than those which precede.

78. Linoleums. — The patterns on ordinary oilcloth are usually printed, but inlaid linoleum is made by glueing pieces of different colors to a canvas backing and welding into one whole by heat and pressure. By this means colors are secured which are proof against wear. It is also clear that pure geometric design may then have a much larger place than in printed linoleums.

79. Mosaics. — The mosaics shown in this section are closely related to the ancient Roman and Pompeian pavements, and for the most part are more elaborate than those related to Byzantine mosaics.

Remains of ancient Roman pavements are found scattered over Europe wherever the Roman dominion extended. In Italy they are composed almost entirely of marble. In countries where marble was scarce they were made of fragments of baked clay, bits of marble, colored glass, and the like. The designs are complicated and beautiful.[1]

In Part I the octagon is shown in three different positions in the given square.

FIRST POSITION OF THE OCTAGON

80. In Figs. 65 to 70 the octagon is shown with all of its vertices located on the sides of the square. When it is regular, they are determined by using the vertices of the given square as centers and the half diagonal as radius. Figures 68 and 69 are the most common of those shown.

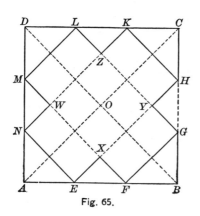

Fig. 65.

Fig. 65*a*. — Parquet Flooring.

81. In Fig. 65 ABCD is a square with BD and AC the diagonals. AN = AE = BF = BG, etc., and the points are joined as indicated.

---

[1] *Enc. Brit.*, Mosaics. Morgan and Zahn (2) give sample pavements.

EXERCISES

1. Prove that NE, MF, and DB are parallel.

2. If $AB = a$ and $AE = \dfrac{a}{n}$, find the area of $EXFGYH$ etc.

$$Ans. \; \frac{2\,a^2}{n^2}(2\,n - 3).$$

3. If $AB = a$ and $AE = \dfrac{a}{3}$, find this same area. $\qquad Ans. \; \dfrac{2\,a^2}{3}.$

4. If $BO = BE$, show that $NE = EF$.

5. If $AB = a$ and $BO = BE$, show that $AE = \dfrac{a}{2}(2 - \sqrt{2})$, and find the area of $EXFGYH$ etc. $\qquad Ans. \; a^2(4\sqrt{2} - 5).$

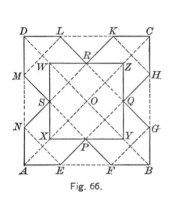

Fig. 66.

Fig. 66a. — Mosaic Flooring. After the Roman.
Barre, Vol. V, 6me Series, Pl. 3.

82. In Fig. 66 ABCD is a square. EFGHKL etc. is a regular octagon inscribed in the square. The points E and H, F and M, etc., are joined. These lines intersect at P, Q, etc. Lines XY, YZ, etc. are drawn through points P, Q, etc., parallel to the sides of the square.

EXERCISES

1. Show how to construct the regular octagon $EFGH$ etc., and give proof.

Suggestion. — See § 29.

2. Prove that $PQRS$ is a square.

3. Prove that $XYZW$ is a square with vertices on the diagonals of the given square.

Suggestion. — Prove that AO is the perpendicular bisector of SP and that $XS = XP$. Therefore X lies on AO.

4. Prove that $AEPX$, $PFBY$, $BGQY$ are congruent rhombuses.

5. If $AB = a$, find the length of AE and EF.

$$Ans. \ (1) \ \frac{a}{2}(2 - \sqrt{2}); \ (2) \ a(\sqrt{2} - 1).$$

6. If $AB = a$, find the length of XY and area of $WXYZ$.

$$Ans. \ (1) \ a(2 - \sqrt{2}); \ (2) \ a^2(6 - 4\sqrt{2}).$$

7. If $AB = a$, find the area of $\triangle EFP$ and of the rhombus $AEPX$.

$$Ans. \ (1) \ \frac{a^2}{4}(3 - 2\sqrt{2}); \ (2) \ \frac{a^2}{4}(3\sqrt{2} - 4).$$

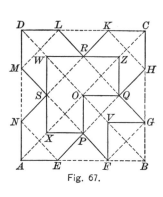

Fig. 67.

83. In Fig. 67 ABCD is a square with diagonals AC and BD.

Fig. 67a. — Mosaic Flooring.

EXERCISES

1. Show how Fig. 67 is obtained from Fig. 66.

2. If $AB = a$, find the area of $POQZRWX$.

$$Ans. \ \frac{a^2}{2}(9 - 6\sqrt{2}).$$

3. If $AB = a$, find the area of the square $BGVF$.

$$Ans. \ \frac{a^2}{2}(3 - 2\sqrt{2}).$$

4. Prove that $GQOV$ and $RZCK$ are congruent rhombuses.

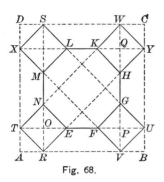

Fig. 68.

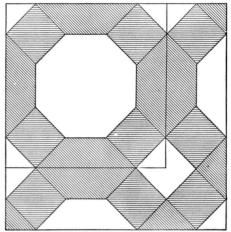

Fig. 68a. — Linoleum Design.

84. In Fig. 68 ABCD is a square, AR = VB = BU, etc., and the points are joined as indicated.

EXERCISES

1. Prove that $TREN$ and $VUGF$ are congruent squares.

2. If $EFGHK$ etc. is a regular octagon, prove that (1) $OE : EF : FP$ $:: 1 : \sqrt{2} : 1$, and (2) $AR : RV : VB :: 1 : 2 + \sqrt{2} : 1$.

3. Construct Fig. 68 in a given square so that $EFGHK$ etc. is a regular octagon.

4. If $AB = a$, find the lengths of AR, RT, and RV, when $EFGH$ etc. is a regular octagon.

$Ans.$ (1) $\frac{a}{14}(4 - \sqrt{2})$; (2) $\frac{a}{7}(2\sqrt{2} - 1)$; (3) $\frac{a}{7}(3 + \sqrt{2})$.

5. If $AB = a$ and if $EFGH$ etc. is a regular octagon, find the following areas:

(a) $\triangle ART$; (b) $TREN$; (c) $RVFE$; (d) $EFGHK$ etc.

$Ans.$ (a) $\frac{a^2}{196}(9 - 4\sqrt{2})$; (b) $\frac{a^2}{49}(9 - 4\sqrt{2})$;

(c) $\frac{a^2}{98}(5\sqrt{2} + 1)$; (d) $\frac{2a^2}{49}(5\sqrt{2} + 1)$.

Note. — For a different discussion of the designs given in Fig. 68a see § 30.

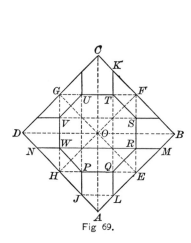

Fig 69.

Fig. 69a. — Linoleum Design.

85. In Fig. 69 ABCD is a square, EFGH the square formed by joining the middle points of the sides, PQRST etc. the regular octagon inscribed in the square EFGH, and the points are joined as shown.

EXERCISES

1. Prove that (*a*) VS and WR are parallel to DB; (*b*) QR is parallel to AB; (*c*) QR is ½ of LM; (*d*) $AL = LQ$; (*e*) PQ is equal and parallel to JL; (*f*) $\dfrac{AL}{LE} = \dfrac{1}{\sqrt{2}}$; (*g*) $LMRQ$ is an isosceles trapezoid.

2. If $AB = a$, find the lengths of AL and LE.

$$Ans. \;\; (1)\; \frac{a}{2}(\sqrt{2}-1); \;\; (2)\; \frac{a}{2}(2-\sqrt{2}).$$

3. If $AB = a$, find the areas of (1) $LMRQ$, (2) $ALQPJ$, and (3) the octagon $PQRST$ etc.

$$Ans. \;\; (1)\; \frac{3\,a^2}{8}(3-2\sqrt{2}); \;\; (2)\; \frac{a^2}{8}(4\sqrt{2}-5); \;\; (3)\; a^2(\sqrt{2}-1).$$

Note.—For a different discussion of the design in Fig. 68*a* see § 31.

86. In Fig. 70 EFGH is the square formed by joining the middle points of the sides of the square ABCD. PQRS etc. is the regular octagon inscribed in the square EFGH. Points P and Ṣ, Q and V, etc. are joined, thus determining point Y and similar points. Points P and U, S and V, etc. are joined, thus determining points N, K, Z, etc. Points L and Z, N and K, etc. are joined.

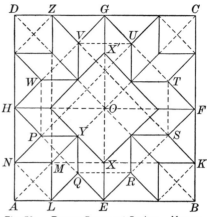

Fig. 70. — Roman Pavement Design. Morgan, p. 204.

EXERCISES

1. Construct a complete figure like Fig. 70.

2. Show the relation between Fig. 70 and Figs. 66 and 69.

3. Prove that (*a*) *ALMN* is a square; (*b*) *YX*, *HE*, and *NL* are parallel; (*c*) $AN = AL = LQ = QE = ER = EX = BK$, etc.; (*d*) *LQYM*, *YQEX*, etc. are congruent rhombuses; (*e*) square $ALMN = \frac{1}{4}$ of square XX'; (*f*) $\dfrac{AL}{LE} = \dfrac{1}{\sqrt{2}}$.

4. If $AB = a$, find the lengths of *AL*, *LE*, and the altitude of the rhombus *MLQY*.

> *Ans.* (1) $\dfrac{a(\sqrt{2}-1)}{2}$; (2) $\dfrac{a}{2}(2-\sqrt{2})$; (3) $\dfrac{a}{4}(2-\sqrt{2})$.

5. If $AB = a$, find the area of one rhombus and of the square *XX'*.

> *Ans.* (1) $\dfrac{a^2}{8}(3\sqrt{2}-4)$; (2) $a^2(3-2\sqrt{2})$.

SECOND POSITION OF THE OCTAGON

87. In Fig. 71 the octagon is inscribed in a given square in such a manner that four of its vertices will coincide with the middle points of the sides of the square.

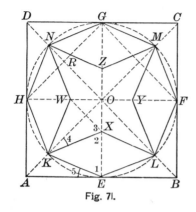

Fig. 7l.

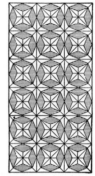

Fig. 7l*a*. — Parquet Floor'ng.
Arabic and Roman.   Calvert
(l), p. 303; Zahn (2), Vol.
2, Pl. 96.

88. In Fig. 71 circle O is tangent to the sides of square ABCD at points
E, F, G, and H, and cuts the diagonals at points K, L, M, and N. Point X on
OE is so chosen that KX = KE. OX = OY = OZ = OW, and the points are
joined as indicated.

EXERCISES

1. Prove that

(*a*) *KELFM* etc. is a regular octagon.

(*b*) *KELX*, *LYMF*, and *MGNZ* are congruent rhombuses.

2. Prove that *KE*, *XL*, *NZ*, and *GM* are parallel.

3. Prove that $XO = AK$.

Suggestion. — Compare $\triangle$ *KXO* and *AEK*. $\angle 4 = \angle 5 = \frac{1}{4}$ rt. $\angle$.

4. If $AB = a$, find lengths of *XO*, *XW*, and *XE*.

$$Ans. \; (1) \; \frac{a}{2}(\sqrt{2}-1); \; (3) \; \frac{a}{2}(2-\sqrt{2}).$$

5. If $AB = a$, find area of the figure *WKXO* and of the entire star
WKXLY etc. $Ans. \; (2) \; \dfrac{a^2}{2}(2-\sqrt{2}).$

6. If $AB = a$, find area of *KXLE* and of entire octagon.

$$Ans. \; (1) \; \frac{a^2}{4}(\sqrt{2}-1); \; (2) \; \frac{a^2\sqrt{2}}{2}.$$

7. Prove that the figure *ELFB* is $\frac{1}{4}$ of the star *KXLYM* etc., and
that the stars shown in Fig. 71*a* are equal.

8. From the results obtained in Exs. 5, 6, and 7, find the propor-
tion of the dark to the light wood in Fig. 71*a*. $Ans. \; \dfrac{1}{\sqrt{2}}.$

9. Join *GH* and *GF*. Prove that *GN* bisects ∠ *DGH*.

10. Construct the square *EFGH*. Prove that a regular octagon is formed by the intersection of the bisectors of ∠ *DGH*, *FGE*, *EFB*, etc.

11. From the theorem, " The bisector of an angle of a triangle divides the opposite side into parts proportional to the other two," find the lengths of *DN* and *NR* and hence of *GN*, if $AB = a$.

$$Ans. \ (1) \ \frac{a}{2}(\sqrt{2} - 1); \quad (2) \ \frac{a}{4}(2 - \sqrt{2}); \quad (3) \ \frac{a}{2}\sqrt{2 - \sqrt{2}}.$$

12. Prove that the points *L*, *X*, and *H* are collinear. Hence, give another method for constructing the figure.

THIRD POSITION OF THE OCTAGON

89. In Figs. 72 and 73 the octagon is concentric with the given square, but is wholly within it, so that none of its vertices lie on the sides of the square.

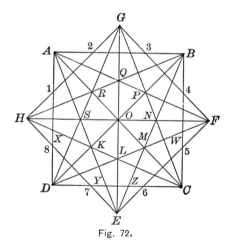

Fig. 72.

EXERCISES

90. 1. Construct Fig. 72 from the square *ABCD*.

Suggestion. — Draw the diameters and diagonals and produce the diameters beyond the sides of the square. Draw *BH* bisecting ∠ *ABD* and extend to meet *OH*. Draw *CH* bisecting ∠ *ACD* and extend to meet *OH*. Continue thus about the square.

Fig. 72a. — Cut-glass Design.

2. Prove that BH, FH, and CH are concurrent.

3. Prove that $OB = OH$, and that $AB = GH$.

4. Prove that $\angle GHE$ is divided into four equal parts by the lines BH, FH, and CH.

5. Prove that lines HB, GD, and AC are concurrent. What other lines are concurrent for the same reason ?

Suggestion. — The bisectors of the angles of a triangle are concurrent.

6. Prove that $BH = EB = GD$, etc.

7. Prove that $KLMNP$ etc. is a regular octagon.

8. Prove that (a) $KX = KY$, (b) $DX = DY$, (c) $DYKX$ is one quarter of a regular octagon.

9. If $AB = a$, find the lengths of OL and DK.

$$Ans.\ (1)\ \frac{a}{2}(2 - \sqrt{2});\ (2)\ a\,(\sqrt{2} - 1).$$

Suggestion. — Use the theorem: " The bisector of an angle of a triangle divides the opposite side into parts proportional to the other two sides."

10. Prove that $A1 = H1$ and that 1–2–3–4–5 etc. is a regular octagon.

11. Prove that (1) DC bisects LE, and (2) $EL = DY = a(\sqrt{2} - 1)$.

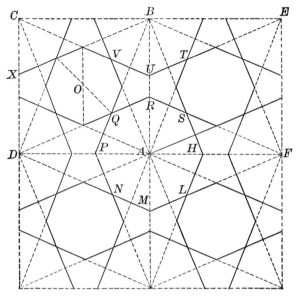

Fig. 73. — Arabic Lattice Design from Cairo.

91. Figure 73 is formed by a repetition of the unit shown in Fig. 72.

EXERCISES

1. Prove that (*a*) EU and BX are parallel; (*b*) EU, CU, and AB are concurrent; (*c*) $TU = VU = RS = RQ$; (*d*) $PNMLH$ etc. is a regular octagon.

2. Prove that the area of octagon O is $\frac{1}{2}$ the area of octagon A and that the side of octagon O is to the side of octagon A as $1 : \sqrt{2}$.

Suggestion. — Consider ratio of OQ to PA. $OQ = \frac{a}{2}(2 - \sqrt{2})$, $PA = a(\sqrt{2} - 1)$, when $DA = a$.

PART 2. DESIGNS CONTAINING EIGHT–POINTED STARS

92. Occurrence. — Star polygons in general are discussed in Part 5 of this chapter. The widespread use of the two eight-pointed stars, both in modern industrial design and in historic ornament, entitles them to separate discussion. The use of many-pointed stars is much more restricted.   Besides the illustrations given in this section, the two eight-pointed stars are very common in medieval mosaics, as in the Duomo de Monreale,[1] the church of San Lorenzo,[2] churches at Ravenna,[3] and elsewhere.[4]

THE THREE-EIGHTHS STAR

93.   The $\frac{3}{8}$ star is formed by joining every third vertex of the octagon.   When inscribed in a given square, it is related to the octagon obtained by cutting off the corners of that square.   (See § 29.)   Both the regular and irregular forms are common.   Two irregular forms of frequent occurrence are discussed in § 94.

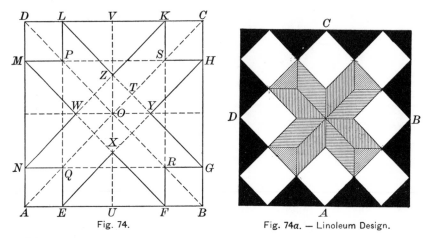

Fig. 74.   Fig. 74a. — Linoleum Design.

94. In Fig. 74 ABCD is a square.   AE = BF = CH, etc.   Points N and G, F and K, etc., are joined; also points E and H, F and M, etc.

EXERCISES

1. Prove that (*a*) *MW* and *ZL* are equal and parallel; (*b*) *PL* and *KS* are equal and parallel; (*c*) *LZ*, *KZ*, and *VO* are concurrent; (*d*) *PL*, *PM*, and *OD* are concurrent; (*e*) *POZL* is a parallelogram.

[1] Gravina, Pl. 8*B*.   [2] Wyatt, Pl. 10.
[3] Gayet, Vol. III, Ravenna, X and XII.   [4] Gruner, Pl. 26.

2. If $AE = \dfrac{AB}{3}$, prove that points Y, S, and Z are collinear. Draw a diagram to illustrate this case.

Suggestion. — Points L, Z, and Y are collinear by construction of the figure. It is necessary to prove that points T and S would coincide. To prove this, notice that if $\dfrac{DL}{DC} = \dfrac{1}{3}$ then, $\dfrac{OT}{OC} = \dfrac{1}{3}$, and if $\dfrac{VK}{VC} = \dfrac{1}{3}$, then $\dfrac{OS}{OC} = \dfrac{1}{3}$.

3. If $AE = \dfrac{AB}{4}$, prove that the points H, S, Z, P, and M, are collinear. Draw a diagram to illustrate this case.

Suggestion. — See Fig. 74*b*.

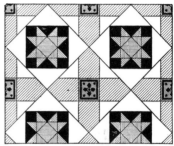

Fig. 74*b*. — Tiled Floor Design. Scale $\frac{3}{4}$ in. = I ft.

4. Show that Fig. 74*a* illustrates both cases $\dfrac{AE}{AB} = \dfrac{1}{3}$ and $\dfrac{AE}{AB} = \dfrac{1}{4}$.

5. If $BO = BE$, prove that

(*a*) $EFGHK$ etc. is a regular octagon, (*b*) $KZ = KS$, (*c*) square $PQRS$ = square $XYZW$.

6. If $AB = a$ and $AE = \dfrac{a}{n}$, find the area of $AEQN$, of $\triangle EXF$, and of the figure $EXFRGY$ etc.

$$\text{Ans. (1) } \frac{a^2}{n^2}; \text{ (2) } \frac{a^2}{4\,n^2}\,(n-2)^2; \text{ (3) } \frac{4\,a^2}{n^2}\,(n-2).$$

7. Find these areas if $AB = a$ and $AE = \dfrac{a}{3}$. $\qquad$ *Ans.* (3) $\dfrac{4\,a^2}{9}$.

8. Find these areas if $AB = a$ and $AE = \dfrac{a}{4}$. $\qquad$ *Ans.* (3) $\dfrac{a^2}{2}$.

9. If $AB = a$ and $AE = \dfrac{a}{n}$, find the area of square $PQRS$ and of square $XYZW$. $\qquad$ *Ans.* (1) $\dfrac{a^2}{n^2}\,(n-2)^2$; (2) $\dfrac{2\,a^2}{n^2}$.

10. If the squares $PQRS$ and $XYZW$ are equal in area, find the value of n when $AB = a$, $AE = \dfrac{a}{n}$. $\qquad$ *Ans.* $n = 2 + \sqrt{2}$.

Suggestion. — Use the equation $\dfrac{2}{n^2} = \dfrac{(n-2)^2}{n^2}$.

11. Draw a diagram to illustrate this case.

12. If $OB = BE$, find the value of n when $AB = a$, $AE = \dfrac{a}{n}$.

Ans. $2 + \sqrt{2}$.

13. If $AB = a$ and $CO = CL$, find the area of the figure $EXFRGY$ etc.

Ans. $2\,a^2\,(3\sqrt{2} - 4)$.

14. If the area of $EXFRGY$ etc. is 64 sq. in., find AB.

Ans. $4\sqrt{3\sqrt{2} + 4}$.

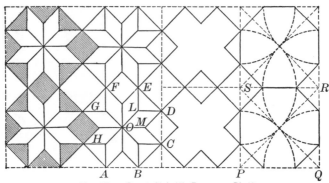

Fig. 75. — Gayet, Vol. III, Ravenna, Pl. 10.

95. Figure 75 is the drawing which shows the construction of the tile design given in Fig. 75a. In Fig. 75 the octagon ABCDE etc. is regular.

EXERCISES

1. If one side of the rhombus-shaped tile $OMDL$ is $\frac{3}{2}\sqrt{2}$, find the length of one side of the octagon $ABCD$ etc., and of the square $PQRS$.

Ans. (1) 3; (2) $3(1 + \sqrt{2})$.

2. If in Fig. 75a each of the small white triangles occupies $\frac{1}{4}$ of the area of the black triangle in which it is placed, find the percentage of black tiling in the whole design.

Ans. About 30 %.

Fig. 75a. — Tiled Floor Design.
Scale $\frac{3}{8}$ in. = 1 ft.

96. In Fig. 76 ABCD is a square. AE = BF = BG, etc. Points E and H, M and F, etc., also L and E, N and G, etc. are joined. Lines EH, FM, etc. intersect the diagonals at points X, Y, Z, and W. Lines XW, WZ, etc. are extended to meet the lines EH, GL, FM, etc., at points S, R, Q, P, etc.

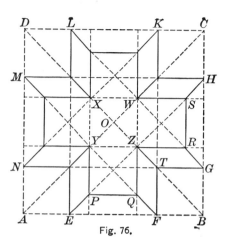

Fig. 76.

EXERCISES

1. Prove that (a) $XYZW$ is a square with YZ parallel to AB; (b) $PQZY$ and $BGTF$ are congruent squares; (c) $ZQFT$ and $ZTGR$ are congruent rhombuses; (d) $EFQP$ is an isosceles trapezoid.

2. If $AB = a$ and $AE = \dfrac{a}{n}$, find the following areas: (a) the cross $YPQZRS$ etc.; (b) the rhombus $FTZQ$; (c) the trapezoid $EFQP$.

$$Ans. \quad (a)\ \frac{5\,a^2}{n^2};\ (b)\ \frac{a^2}{2\,n^2}(n-3);\ (c)\ \frac{a^2}{4\,n^2}(n-1)(n-3).$$

3. If $AE = EP = PQ = QF = FB = BG$, etc., prove that $AE : EF : FB : : 1 : 1 + \sqrt{2} : 1$.

4. Construct a figure to illustrate the relations given in Ex. 3. (See Fig. 76a and 76b.)

5. If the relations found in Ex. 3 hold true, prove that

$$AE = \frac{a}{7}(3 - \sqrt{2}).$$

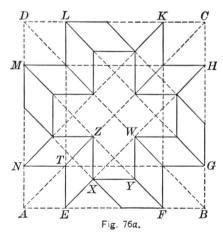

Fig. 76a.

Fig. 76b. — Mosaic Floor Design.

6. If $AB = a$ and $AE = \dfrac{a(3 - \sqrt{2})}{7}$, find the area of the square $AETN$, of the rhombus $EXZT$, and of the entire cross in Fig. 76a.

Ans. (1) $\dfrac{a^2}{49}(11 - 6\sqrt{2})$; (2) $\dfrac{a^2}{98}(11\sqrt{2} - 12)$; (3) $\dfrac{5\,a^2}{49}(11 - 6\sqrt{2})$.

THE TWO-EIGHTHS STAR

97. The $\frac{2}{8}$ star is formed by joining every second vertex of the octagon and is composed of two superposed squares. When inscribed in a given square, it is related to the octagon discussed in § 88. Figures 77 and 83 show it as it most commonly occurs. Fitted together with certain peculiar crosses, as in Fig. 77a, it forms a very common all-over pattern, which was executed in tiles in Saracenic decoration.[1] Figure 78 is an Arabic form. The irregular form suggested in § 99, Ex. 4, is from a Roman mosaic.[2] Figure 83 occurs also in both the Arabic and the Roman, and is extremely common in modern ornament.

[1] Furnival Pl. 81, Cut.   [2] Price.

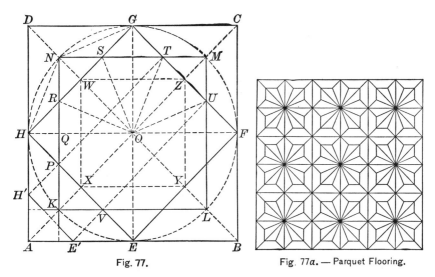

Fig. 77.

Fig. 77a. — Parquet Flooring.

98. In **Fig. 77** circle O is tangent to the sides of the square ABCD at points E, F, G, and H, and intersects the diagonals at points K, L, M, and N Points E, F, G, and H, also K, L, M, and N are joined. Then the other points are joined as indicated.

EXERCISES

1. Prove that

(*a*) *EFGH* and *KLMN* are congruent squares; (*b*) *XYZW* is a square; (*c*) *DA*, *KN*, and *XW* are parallel; (*d*) *HR = RN = NS*; (*e*) *RO* bisects $\angle NOH$; (*f*) *PRSTU* etc. is a regular octagon.

2. If $AB = a$, find area of circle that could be inscribed in octagon *PRSTU* etc. *Ans.* $\dfrac{\pi a^2}{8}$.

3. If $AB = a$, find the length of *HR* and *RS*.

Ans. (1) $\dfrac{a}{2}(\sqrt{2} - 1)$; (2) $\dfrac{a}{2}(2 - \sqrt{2})$.

4. If $AB = a$, find the area of figure *OSNR* and of the entire star *KPHRN* etc. *Ans.* (1) $\dfrac{a^2}{8}(2 - \sqrt{2})$; (2) $a^2(2 - \sqrt{2})$.

5. Compare results in Ex. 4 with those given for Fig. 71*a*. (See § 88, Ex. 5.) Notice that the area of star *KPHRN* etc. (Fig. 77) is twice the area of star *WKXLY* etc. in Fig. 71.

6. Find area of figure *HRNSGD*. $Ans. \dfrac{a^2}{4}(\sqrt{2} - 1).$

7. From results of Exs. 4 and 6 compare the areas of the stars and crosses in Fig. 77a. $Ans. \dfrac{\text{area cross}}{\text{area star}} = \dfrac{1}{\sqrt{2}}.$

8. If *GH = a*, find lengths of *HR*, *RS*, and *AB*.

$Ans.$ (1) $\dfrac{a}{2}(2 - \sqrt{2})$; (2) $a(\sqrt{2} - 1)$; (3) $a\sqrt{2}.$

9. If *GH = a*, find the area of *ORNS* and of the entire star *KPHRNSG* etc. $Ans.$ (1) $\dfrac{a^2}{4}(2 - \sqrt{2})$; (2) $2\,a^2(2 - \sqrt{2}).$

10. In Fig. 77 join *P* and *T*, also *U* and *V*, and extend to meet *AD* and *AB*. Extend *NK* and *LK* to meet *AD* and *AB*. Prove that

(*a*) *AK = KP = AH′.*

(*b*) The figure *AH′KE′* is ¹ of a star similar to the star *KPHRNS* etc.

11. If *AB = a*, find the area of the figure *AH′KE′*.

$Ans. \dfrac{a^2(10 - 7\sqrt{2})}{4}.$

99. Figure 78 exhibits the star shown in Fig. 77 inscribed in a given square with all of its vertices on the sides of the square.

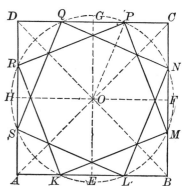

Fig. 78. — From Ceiling of the Infanta's Tower in the Alhambra. Calvert (I), p. 34I.

EXERCISES

1. Show how to construct Fig. 78.

Suggestion. — The vertices of the star are determined by using the vertices of the square as centers and the half diagonal as radius.

2. If $AB = a$, find the length of (1) KM and (2) the radius of the circumscribed circle.   *Ans.* (1) $a\sqrt{2 - \sqrt{2}}$; (2) $\dfrac{a}{2}\sqrt{4 - 2\sqrt{2}}$.

3. If $AB = a$, find the area of the star.   *Ans.* $2\,a^2\,(2 - \sqrt{2})^2$.

4. Construct a figure in which points K, L, M, N, etc. shall be the middle points of AE, EB, BF, FC, etc.

5. Show that a circle can be circumscribed about the star constructed as in Ex. 4, and find its radius, if $AB = a$.   *Ans.* $\dfrac{a}{4}\sqrt{5}$.

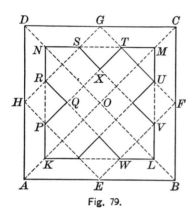

Fig. 79.

Fig. 79a. — Parquet Flooring.

100. In Fig. 79 ABCD is a square.   E, F, G, and H the middle points of the sides.   OK = OL = OM = ON.   The squares EFGH and KLMN are drawn, and the points are joined as indicated.

EXERCISES

1. Prove that $HPQR$ and $GSXT$ are equal squares.

2. Prove that $AK = PH$.

3. Prove that $\dfrac{AK}{AO} = \dfrac{RP}{AD}$, and $\dfrac{KO}{AO} = \dfrac{KL}{AB}$.

4. If $AB = a$ and $AK = \dfrac{AO}{n}$, find the area of (1) $\triangle PQR$, (2) $KPQRNS$ etc., (3) square $KLMN$.

Ans. (1) $\dfrac{a^2}{4\,n^2}$; (2) $\dfrac{a^2}{n}(n - 2)$; (3) $\dfrac{a^2}{n^2}(n - 1)^2$.

5. If the area of *KPQRNS* etc. is $\frac{1}{2}$ that of the square *ABCD,* find *AK.*   *Ans.* $AK = \dfrac{AO}{4}.$

Suggestion. — Use equation $\dfrac{a^2}{n}(n-2) = \dfrac{a^2}{2}.$

6. If the area of the cross is $\dfrac{1}{k}$ that of the square, find *n* and *AK.*

Ans. $n = \dfrac{2\,k}{k-1}.$

7. If $OG = ON$ and the squares *EFGH* and *KLMN* are equal, as in Fig. 77, find the area of $\triangle PQR$ and of the cross *KPQRNS* etc., if $AB = a.$

Ans. (1) $\dfrac{a^2}{8}(3 - 2\sqrt{2})$; (2) $a^2(\sqrt{2} - 1).$

Suggestion. — From § 98, Ex. 3, $HR = \dfrac{a}{2}(\sqrt{2} - 1)$ and $RS = RP$

$= \dfrac{a}{2}(2 - \sqrt{2}).$

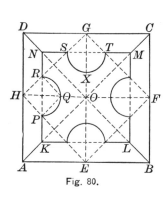

Fig. 80.

Fig. 80*a.* — Steel
Ceiling Design.

101. In Fig. 80 ABCD is a square.   E, F, G, and H are the middle points of the sides.   OK = OL = OM = ON.   The squares EFGH and KLMN are drawn, and the points are joined as indicated.   Semicircles are constructed on the lines PR, ST, etc. as diameters.

EXERCISES

1. If $AB = a$ and $AK = \dfrac{AO}{n}$, find the area of the semicircle *PQR.*

Ans. $\dfrac{\pi a^2}{2\,n^2}.$

2. If $AB = a$ and $AK = \dfrac{AO}{n}$, find the area bounded by PK, $\overset{\frown}{PQR}$, RN, NS, $\overset{\frown}{SXT}$, etc.

$$Ans. \ \frac{a^2}{n^2}\,[(n-1)^2 - 2\,\pi].$$

3. If $AB = a$ and $AK = \dfrac{AO}{n}$, find the area bounded by $\overset{\frown}{PQR}$ and line-segments RN, ND, DA, AK, and KP.

$$Ans. \ \frac{a^2}{4\,n^2}\,(2\,n - 1 + 2\,\pi).$$

4. If $AB = a$ and $OG = ON$, find the area bounded by $\overset{\frown}{PQR}$ and the line-segments RN, NS, etc. $\qquad Ans. \ \dfrac{a^2}{8}\,[4 - \pi(6 - 4\sqrt{2})].$

Suggestion. — From § 98, Ex. 3, $RP = SR = \dfrac{a}{2}\,(2 - \sqrt{2}).$

5. If $AB = a$ and $OG = ON$, find the area bounded by AK, KP, $\overset{\frown}{PQR}$, RN, ND, AD. $\qquad Ans. \ \dfrac{a^2}{32}\,[4 + \pi(6 - 4\sqrt{2})].$

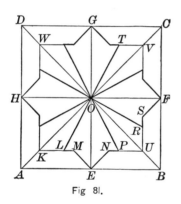

Fig 8l.

Fig. 8l*a.* — Parquet Flooring.

102. In Fig. 81 ABCD is a square. E, F, G, and H are the middle points of the sides. OK = OU = OV = OW. The squares EFGH and KUVW are drawn. KL = UP = UR = VT, etc., and the points are joined as indicated.

EXERCISES

1. Prove that (*a*) $KM = NU = US$, etc.; (*b*) $EN = FS$, etc.; (*c*) $AEMK$ and $EBUN$ are congruent trapezoids; (*d*) L, O, and T are collinear.

2. If $OK = OE$, prove that $EN = NU$.

3. If $AB = a$, $AK = \dfrac{AO}{6}$, and $KL = LM$, find the areas of (1) $KUVW$; (2) $AEMK$; (3) $LMENPO$.

4. Find these same areas if $AK = \dfrac{AO}{n}$, $AB = a$, and $KL = m \cdot LM$.

Ans. (1) $\dfrac{a^2}{n^2}(n-1)^2$; (2) $\dfrac{a^2}{4n^2}(n-1)$; (3) $\dfrac{a^2}{4n^2}\left(\dfrac{mn + n^2 - 2n + 2}{m + 1}\right)$.

5. If $AB = a$ and $AK = \dfrac{AO}{6}$, what is the length of KL, if $LMENPO$ occupies $\frac{1}{8}$ of the square.   *Ans.* $KL = \frac{2}{3}LM$.

Suggestion. — Use the equation

$$\frac{a^2}{4n^2}\left(\frac{mn + n^2 - 2n + 2}{m + 1}\right) = \frac{a^2}{8} \text{ and substitute } n = 6.$$

6. If $LM = 0$ and $AK = \dfrac{AO}{6}$, what part of the square is occupied by $LMENPO$?   *Ans.* $\frac{1}{24}$.

7. If $KL = LM$ and $AK = \dfrac{AO}{n}$, what part of the square is occupied by this figure?   *Ans.* $\dfrac{n^2 - n + 2}{8n^2}$.

DESIGNS COMBINING THE THREE-EIGHTHS AND THE TWO-EIGHTHS STARS

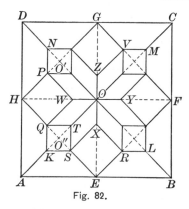

Fig. 82.

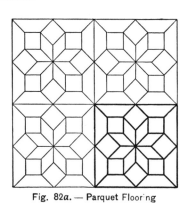

Fig. 82a. — Parquet Flooring

103. In Fig. 82 ABCD is a square. E, F, G, and H are the middle points of the sides. OK = OL = OM = ON. The squares EFGH and KLMN are drawn. Points Q and V, P and R, etc., also K and L, V and R, K and N, etc., are joined.

EXERCISES

1. Prove that (*a*) *HQWP*, *KSTQ*, and *XYZW* are squares; (*b*) *QTWO* is a parallelogram.

2. If $AB = a$, and $O''K = \dfrac{O''A}{n}$, find the areas of the squares *KSTQ* and *HQWP*.

$$Ans. \ (1) \ \frac{a^2}{4\,n^2}; \ (2) \ \frac{a^2}{8\,n^2}(n-1)^2.$$

3. Find the area of the star *HQKSE* etc. if $AB = a$, and $O''K = \dfrac{O''A}{n}$.

$$Ans. \ \frac{a^2}{2\,n^2}(n^2+1).$$

4. Also find the area of the star *PWQTSX* etc.

$$Ans. \ \frac{a^2}{n^2}(n-1).$$

5. Also find the area of the trapezoid *AESK*.

$$Ans. \ \frac{a^2}{16\,n^2}(n^2-1).$$

6. If $ON = OG$ and $AB = a$, find the area of star *PWQTSX* etc.

$$Ans. \ a^2(3\sqrt{2}-4).$$

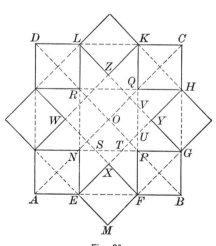

Fig. 83.

Fig. 83*a*. —Oilcloth Design.

104. In Fig. 83 ABCD is a square. AE = BF = BG = CH, etc. Points are joined as indicated.

EXERCISES

1. Prove that $MFXE$, $NPQR$, and $XYZW$ are squares.

2. If $AB = a$ and $AE = \dfrac{a}{3}$, find the area of

(1) $EMFX$; (2) $XYZW$; (3) $AEMFBG$ etc.; (4) $EFGHK$ etc.; (5) $AEXFBG$ etc.; (6) $NEMFPG$ etc.

Ans. (1) $\dfrac{a^2}{18}$; (2) $\dfrac{2\,a^2}{9}$; (3) $\dfrac{10\,a^2}{9}$; (4) $\dfrac{7\,a^2}{9}$; (5) $\dfrac{8\,a^2}{9}$; (6) $\dfrac{2\,a^2}{3}$.

3. Draw a figure to illustrate the case

$$AE = \frac{AB}{3}$$

and prove that W, N, and X are collinear. (See § 94, Ex. 2.)

4. If $AB = a$

and $\quad AE = \dfrac{a}{4}$,

find the length of EM.

Ans $\dfrac{a}{4}\sqrt{2}$.

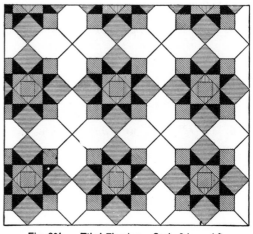

Fig. 83*b*. — Tiled Flooring.   Scale ¾ in. = I ft.

5. If $AB = a$ and $AE = \dfrac{a}{4}$, find the area of (1) $AEMFBG$ etc.; (2) $EFGHK$ etc.; (3) $AEXFBGY$ etc.; (4) $NEMFPG$ etc.

Ans. (1) $\dfrac{5\,a^2}{4}$; (2) $\dfrac{7\,a^2}{8}$; (3) $\dfrac{3\,a^2}{4}$; (4) a^2.

6. If $AB = a$ and $AE = \dfrac{a}{n}$, find the above areas.

Ans. (1) $\dfrac{2\,a^2}{n^2}(n^2 - 2\,n + 2)$; (2) $\dfrac{a^2}{n^2}(n^2 - 2)$; (3) $\dfrac{4\,a^2}{n^2}(n - 1)$;

(4) $\dfrac{2\,a^2}{n}(n - 2)$.

7. If $AF = AO = BE = BH$, etc., prove that $EFGHK$ etc. is a regular octagon and that the squares $WXYZ$ and $NPQR$ are equal.

8. If $AF = AO = BE = BH$, etc. prove that $AE = EM = MF$, etc.

9. If $AB = a$ and $AF = AO$, find the area of $FBGP$.

$$Ans. \ \frac{a^2}{2}(3 - 2\sqrt{2}).$$

10. If $AB = a$ and $AF = AO$, find the four areas of Ex. 5, and also that of $NEXFPGY$ etc.

Ans. (1) $2\,a^2(2 - \sqrt{2})$; (2) $2\,a^2(\sqrt{2} - 1)$; (3) $2\,a^2(\sqrt{2} - 1)$; (4) $2\,a^2(\sqrt{2} - 1)$; (5) $a^2(6\sqrt{2} - 8)$.

11. If $AB = a$ and $AO = AF$, find the length of NS and the area of $NSXTPU$ etc.   *Ans.* (1) $\frac{a}{2}(3\sqrt{2} - 4)$; (2) $a^2(20 - 14\sqrt{2})$.

12. Verify the areas $AEMFB$ etc., and $NSXTPU$ etc. by applying the theorem: "The areas of two similar figures are to each other as the squares on any two homologous sides."

Suggestion. $\dfrac{\dfrac{a^2}{4}(2 - \sqrt{2})^2}{\dfrac{a^2}{4}(3\sqrt{2} - 4)^2} = \dfrac{a^2(4 - 2\sqrt{2})}{a^2(20 - 14\sqrt{2})}.$

13. If $AB = a$ and $AF = AO$, find the area of $ENSX$.

$$Ans. \ \frac{a^2}{2}(5\sqrt{2} - 7).$$

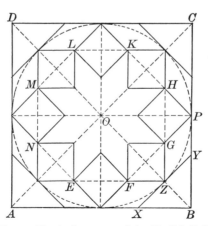

Fig. 84. — From a Roman Mosaic Pavement. Zahn (2), Vol. II, Pl. 79, Vol. III, Pl. 6.

105. 1. Give the construction for the figure within the square *ABCD* in Fig. 84.

Suggestion. — First inscribe the circle in the square.

2. Prove that *GPYZ* is a rhombus.

3. If $OP = r$, find the length of *HK*. *Ans.* $r(2 - \sqrt{2})$.

4. Also find the areas of *ZYPG* and *EFGHK* etc.

$$\textit{Ans. (1) } \frac{r^2}{2}(3\sqrt{2} - 4)\,;\; (2)\; 4\,v^2(\sqrt{2} - 1).$$

PART 3. DESIGNS FORMED FROM ARCS OF CIRCLES

106. Occurrence. — The figures in this section are found in all lines of modern industrial design from all-over lace to iron grills and stamped steel ceilings. In the history of ornament they occur in ancient Assyrian [1] and Egyptian [2], in Chinese [3] and Japanese, in Roman and medieval, and in modern art.

FIGURES FORMED OF SEMICIRCLES AND QUADRANTS

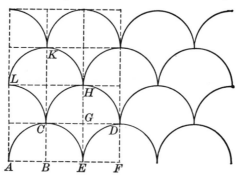

Fig. 85. — Design for Iron Grills and Mosaics,

107. 1. Show how Fig. 85 is constructed from a network of squares.

[1] Ward (1), Vol. I, p. 126. [2] Ward (2), p. 34; Glazier, p. 4.
[3] Clifford, pp. 27–35; Jones (2), Pl. LIX.

2. Prove that the curved figures $CEDHC$ and $CHKLC$ are congruent.

Suggestion. — Draw the chords CE, DE, etc., and CH, HK, etc. Prove that (1) the quadrilaterals thus formed are congruent; (2) corresponding arcs as CE and CH, ED and CH are equal; (3) corresponding arcs lie on the same side of corresponding chords.

3. If $AB = 4$, find the perimeter and area of the curved figure $CEDHC$.

4. If $AB = a$, find the perimeter and area of the curved figure $CEDHC$. *Ans.* (1) $2\pi a$; (2) $2a^2$.

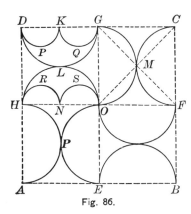

Fig. 86.

Fig. 86a. — Roman Mosaic; also Byzantine. See Gayet, Vol. II, Pl. 30.

108. Figure 86 is constructed on a network of squares. The semicircles are constructed on the sides of the squares as indicated.

EXERCISES

1. Prove that two semicircles are tangent at the intersection of the diagonals of each square.

2. Prove that the curved figures $DPKQGLD$ and $HRNSOLH$ are congruent.

3. If $HO = a$, find the area of the following curved figures: (*a*) $DPKQGLD$; (*b*) $LGMOL$; (*c*) $HRNSOPH$.

Ans. (*a*) $\dfrac{a^2\pi}{16}$; (*b*) $\dfrac{a^2}{2}$; (*c*) $\dfrac{a^2}{16}(8 - \pi)$.

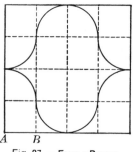

Fig. 87. — From a Roman Pavement.

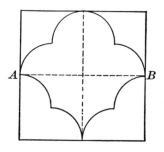

Fig. 88. — Modern Tile. Arabic Design. Prisse d'Avennes, Vol. II, Pl. Cl.

EXERCISES

109. 1. Give the construction for the design unit shown in Fig. 87.

2. If $AB = a$, find the area and the perimeter of the curved figure.

Ans. $8\, a^2$, $4\, \pi a$.

EXERCISES

110. 1. Give the construction for Fig. 88.

2. If $AB = a$, find the area and the perimeter of the curved figure.

Ans. πa, $\dfrac{a^2}{2}$.

OVALS FORMED OF INTERSECTING QUADRANTS

111. Occurrence. — While the illustrations that accompany the following figures are all from modern sources, these forms abound in Roman, Byzantine, and Gothic work. Some instances are especially noteworthy. In the old Singing School in Worcester Cathedral [1] and in the Baptistry at Florence [2] are pavements containing a large number of these designs. In the Basilica Antoniana at Padua is a stone parapet containing numerous different designs based on circles and squares. Figures 89 and 90, as well as trefoils, quadrifoils, six- and eight-pointed rosettes, are

[1] Shaw (1), Pl. xxx. [2] Waring (1), Pl. 23 and 25.

among them.[1] These figures are found in churches and cathedrals in Ravenna, Parenzo,[2] Venice, and many other places.   Figure 90 is found in the details of a window in the Milan Cathedral.[3]   Figure 92 is a common Gothic design used in tracery windows and inlaid tiles.[4] See Fig. 217.

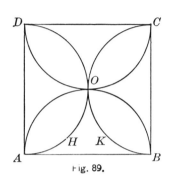

Fig. 89.

112. In Fig. 89 ABCD is a square. The semicircles are constructed within the square on the sides as diameters, as shown in the figure.

EXERCISES

1. Prove that the four leaves formed are congruent.

2. If $AB = 12$, find the area of each leaf and of the curved figure $AHOKB$.

3. Find these same areas if $AB = a$.

$$Ans. \ (1) \ \frac{a^2}{8}(\pi - 2); \ (2) \ \frac{a^2}{8}(4 - \pi).$$

4. If the area of each leaf is 25, find AB.   *Ans.* 13.2[+].

5. If the area of each leaf is s, find AB. $Ans. \ \sqrt{\dfrac{8\,s}{\pi - 2}}.$

6. Show that this figure has four axes of symmetry and also a center of symmetry.

---

[1] Arte Italiana, 1896, Pl. 47.

[2] Gayet, Vol. III ; Ravenna, Pl. 10 ; Vol. II, Parenzo, Pl. 24.

[3] See § 204, Ex. 4.   [4] Shaw (1), Pl. xxx.

     Hartung, Vol. II, Pl. 96.

113. 1. Give the construction for Fig. 90 and prove that the ovals are congruent.

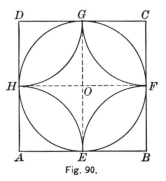

2. If $AB = 12$, find the area of the four-pointed figure in the center and of one of the ovals.

3. If $AB = a$, find these same areas.

Ans. (1) $\dfrac{a^2}{4}(4 - \pi)$; (2) $\dfrac{a^2}{8}(\pi - 2)$.

Fig. 90.

4. If the area of the central figure is s, find AB. 　　*Ans.* $2\sqrt{\dfrac{s}{4 - \pi}}$.

5. What per cent of the area of the square is the area of the central figure? 　　*Ans.* 21.4 % nearly.

Fig. 90*a*. — Iron Grills.

114. 1. Draw a network of squares. Draw the diagonals of each square. Prove that these diagonals form two sets of parallel lines that intersect in squares. See Fig. 91.

2. Draw the diameters of the first set of squares. Compare the areas of squares *ABCD*, *ECFD*, and *AGEH*.

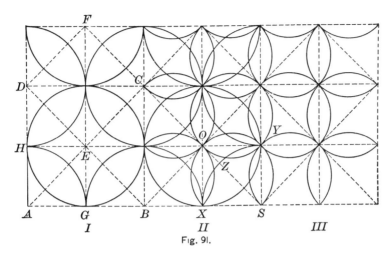

Fig. 91.

3. If the circles in column III, Fig. 91, are circumscribed about the small squares, find the area of one of the ovals formed if $AB = a$.

$$Ans. \ \frac{a^2}{16}(\pi - 2).$$

4. In column III, prove that the diagonals of the large squares are tangent to two circles at the vertices of the small squares.

5. Show how the ovals are formed in column I. If $AB = a$, find the area of each oval.

$$Ans. \ \frac{a^2}{8}(\pi - 2).$$

6. Compare the area of one of the ovals in column III with the area of one in column I. Explain the relation found.

7. In column I, prove that the diameters of each of the large squares are tangent to two circles at the center of each side of the square.

8. If $AB = a$, find the area of the crescent $XZYOX$.

$$Ans. \ \frac{a^2}{8}.$$

9. Draw a diagram showing the construction of Fig. 91a.

Suggestion. — See column II, Fig. 91,

Fig. 91a. — Steel Ceiling Design.

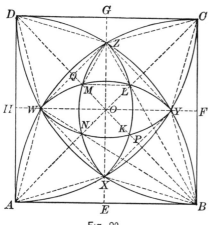

Fig. 92.

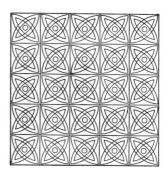

Fig. 92a. — Tiled Flooring. Gothic.
Shaw, (I), Pl. XXX.

EXERCISES

115. 1. Construct Fig. 92, given the square *ABCD*.

Suggestion. The arcs are drawn with the vertices *A*, *B*, *C*, and *D* as centers and intersect in the points *X*, *Y*, *Z*, and *W*. The inner figure is similarly drawn from points *X*, *Y*, *Z*, and *W*.

2. Prove that

(*a*) The points *X*, *Y*, *Z*, and *W* lie on the diameters of the square *ABCD*.

(*b*) The lines *WB* and *ZB* trisect the angle *ABC*.

(*c*) The chords *AW*, *WZ*, and *ZC* are equal.

(*d*) *XBY* is an equilateral triangle.

(*e*) *XYZW* is a square.

(*f*) Line *AC* is the perpendicular bisector of *WX* and *ZY* and bisects ∠*ZCY*.

(*g*) Lines *BY* and *BZ* trisect ∠*CBD*.

(*h*) The curved triangles *BXY* and *WAX* are congruent.

3. In the square *XYZW* construct arcs similar to those constructed in the square *ABCD*. Prove that *ZLM* is an equilateral triangle and that the points *Z*, *L*, *B* are collinear.

4. Construct in a given circle a figure like the one here shown in a square.

5. If $AB = a$, find the length of OP, PB, XY, and BQ.

Ans. (1) $\dfrac{a}{4}(\sqrt{6} - \sqrt{2})$; (2) $\dfrac{a}{4}(3\sqrt{2} - \sqrt{6})$; (3) $\dfrac{a}{2}(\sqrt{6} - \sqrt{2})$;

(4) $\dfrac{a}{4}(\sqrt{6} + \sqrt{2})$.

Suggestion. — Since BXY is an equilateral triangle and $XYZW$ is a square, OB is divided at point P in the ratio $\dfrac{1}{\sqrt{3}}$.

6. If $AB = a$, find the area of the following figures:

(*a*) The linear figure $XYZW$.   *Ans.* $a^2(2 - \sqrt{3})$.

(*b*) The linear $\triangle XBY$.   *Ans.* $\dfrac{a^2}{4}(2\sqrt{3} - 3)$.

(*c*) The linear $\triangle BWZ$.   *Ans.* $\dfrac{a^2}{4}$.

(*d*) The segment bounded by the arc WZ and the chord WZ.

$$Ans.\ \dfrac{a^2}{12}(\pi - 3).$$

(*e*) The curved figure $XYZW$.   *Ans.* $a^2\left(1 - \sqrt{3} + \dfrac{\pi}{3}\right)$.

(*f*) The curved $\triangle BXY$.   *Ans.* $\dfrac{a^2}{4}\left(2\sqrt{3} - 4 + \dfrac{\pi}{3}\right)$.

(*g*) The four-pointed curved figure $AXBYCZ$ etc.

$$Ans.\ a^2\left(\sqrt{3} + \dfrac{2\pi}{3} - 3\right).$$

(*h*) The curved figure AXB.   *Ans.* $\dfrac{a^2}{4}\left(4 - \sqrt{3} - \dfrac{2\pi}{3}\right)$.

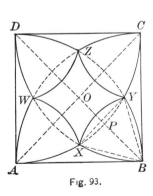

Fig. 93.

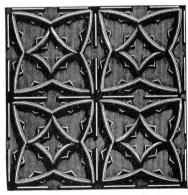

Fig. 93a. — Steel Ceiling Design.

116. **1.** Given the square $ABCD$, construct Fig. 93.

2. Using $XY = \dfrac{a}{2}(\sqrt{6} - \sqrt{2})$ as found in § 115, Ex. 5, find the area of the following figures if $AB = a$:

(*a*) The segment bounded by $\overparen{XY}$ and the chord XY.

$$Ans.\ \frac{a^2}{12}(2\,\pi - 3\sqrt{3})(2 - \sqrt{3}).$$

(*b*) The curved figure $XYZW$. $Ans.\ \dfrac{a^2}{3}(2-\sqrt{3})(3-2\,\pi+3\sqrt{3}).$

(*c*) The curved figure BXY. $Ans.\ \dfrac{a^2}{6}(3\,\pi - \pi\sqrt{3} - 3).$

3. Construct in a given circle a figure like Fig. 93.

CUSPED QUADRILATERALS

117. Occurrence. — The cusped quadrilateral shown in Figs. 94 and 95 is the basis of a large number of fan vaulting designs besides those here mentioned. See also Figs. 145 and 146. This type of vaulting is peculiar to the late English or Perpendicular Gothic architecture. The name refers to the manner in which the main ribs radiate from the tops of columns forming inverted cones.[1]

118. In Fig. 94 the arcs HPE EQF, etc. are drawn with the vertices of the squares as centers and the half side as radius. The five circles are tangent as shown.

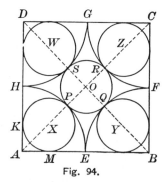

Fig. 94.

[1] Fletcher, p. 288–290; Bond, chapter xxii; Fergusson, Vol. II, p. 361.

EXERCISES

1. Prove that the sectors HAE, EBF, etc., are congruent.

2. Show how to construct (1) circle X inscribed in the sector HAE, and (2) circle O inscribed in the curved figure $EFGH$.

3. Prove that the diagonals AC and BD pass through the points of tangency P, Q, R, and S, and that circles X and O are tangent.

4. If $AB = a$, find the lengths of OP and PX.

$$Ans.\ OP = PX = \frac{a}{2}(\sqrt{2} - 1).$$

5. If $AB = a$, find the area of one of the small circles.

$$Ans.\ \frac{a^2\pi}{4}(3 - 2\sqrt{2}).$$

6. If the area of one of the small circles is s, find AB.

$$Ans.\ \frac{2}{\sqrt{2} - 1}\sqrt{\frac{s}{\pi}}.$$

7. What per cent of the area of the square is occupied by the five circles? $Ans.$ 67 % about.

8. If $AB = a$, find the area of the curved figures (a) $HEFG$; (b) PEQ; (c) EPM.

$$Ans.\ (a)\ \frac{a^2}{4}(4 - \pi);\ (b)\ \frac{a^2}{8}(2 - 2\pi + \pi\sqrt{2});$$

$$(c)\ \frac{a^2}{16}(4\sqrt{2} - 4\pi + 3\pi\sqrt{2} - 6).$$

9. Show that AO is divided by the points P and X in the ratio $\sqrt{2}:1:1$.

10. What per cent of the curved figure $EFGH$ is occupied by the small circle $PQRS$? $Ans.$ 62.8 % about.

Remark. — Details of this figure occur in fan vaulting in Ely Cathedral (Brochure Series, 1901, p. 240) and in Pompeian Mosaic (Zahn (2), Vol. II, Pl. 56).

119. In Fig. 95, ABCD is a square. The arcs HPE, EQF, etc., are drawn with A, B, C, and D as centers and the half side as radius. They intersect the diagonals of the square at points P, Q, R, and S. The circles X, Y, Z, and W are inscribed in the curved figures, PEQ, QFR, etc.

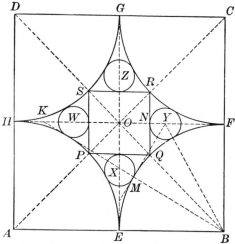

Fig. 95.　From Fan Vaulting. Henry VII Chapel, Westminster Abbey. A. Pugin, Frontispiece.

EXERCISES

1. Prove that $PQRS$ is a square.

2. If $AB = a$, find the area of $PQRS$.　　*Ans.* $\dfrac{a^2}{2}(3 - 2\sqrt{2})$.

3. Construct the small circles X, Y, Z, and W.

Suggestion. — To obtain point Y, make $NK = BQ$. Find point Y on line HF equally distant from K and B. Prove that Y will be the center of a circle tangent to EQF and to the line RQ at N.

4. If $AB = a$, find the length of NF, FK, and KB.

Ans. (1) $\dfrac{a}{4}\sqrt{2}$; (2) $\dfrac{a}{4}(2 + \sqrt{2})$; (3) $\dfrac{a}{4}\sqrt{10 + 4\sqrt{2}}$.

5. Find the per cent of error if we say $KB = AB$.
　　　　　　　　　　　　　　　　Ans. 1.2% nearly.

6. If $AB = a$, find the length of $NY(= x)$.　　*Ans.* $\dfrac{a}{8}(2 - \sqrt{2})$.

Suggestion. — $BY = \dfrac{a}{2} + x$;　$YF = \dfrac{a}{4}\sqrt{2} - x$;　$\overline{BY}^2 - \overline{YF}^2 = \overline{FB}^2$.

Therefore get the equation $\left(\dfrac{a}{2} + x\right)^2 - \left(\dfrac{a\sqrt{2}}{4} - x\right)^2 = \dfrac{a^2}{4}$.

7. Construct a circle which shall be tangent to a given circle and to a given line at a given point.

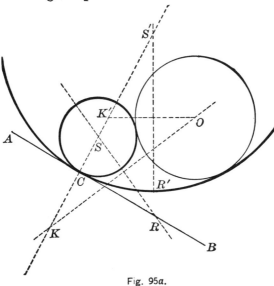

Fig. 95a.

Suggestion. — See Fig. 95a. Here the problem is to construct a circle tangent to circle O and to the line AB at point C. Use method suggested in Ex. 3 above. Show that two circles are possible. For another method of solving this problem see § 246.

120. In Fig. 96, ABCD is a square with the inscribed circle. The arcs AXB, BYC, etc., are tangent to each other at the points A, B, C, and D. The lines KL, LM, etc., are tangent to $\overset{\frown}{AXB}$, $\overset{\frown}{BYC}$, etc., at the points X, Y, etc. Points K, L, M, and N are the centers and KO is the radius for the intersecting quadrants shown.

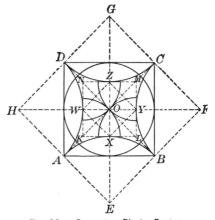

Fig. 96. — Decorative Plaster Design.

EXERCISES

1. Show how to construct $\overset{\frown}{AXB}$, $\overset{\frown}{BYC}$, etc.

2. If $AB = a$, find the area of the curved figure $AXBYCZ$ etc.

$$Ans. \ \frac{a^2}{2}(4 - \pi).$$

3. Prove that $KLMN$ is a square and find its area if $AB = a$.

$$Ans. \ 2\,a^2(3 - 2\sqrt{2}).$$

4. Prove that $XYZW$ is a square and find its area if $AB = 6$.

$$Ans. \ 6.2 \text{ nearly.}$$

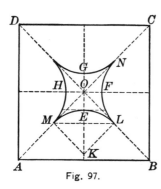

Fig. 97.

Fig. 97*a*. — Steel Ceiling Design.

121. In Fig. 97 ABCD is a square with its diagonals and diameters. OE = OF = OG = OH. This figure shows the cusped quadrilateral formed of arcs that are not necessarily quadrants as they are in Figs. 94–96.

EXERCISES

1. Construct the arc MEL tangent to OA and OB and passing through point E.

Suggestion. — Use Fig. 97*b*. The problem is to construct a circle tangent to AB and AC and passing through point P in the bisector of $\angle BAC$.

2. In Fig. 97b:

(a) If $AP = 8$ and $EP = 6$, find $BO\ (= x)$.

Ans. 12.

Suggestion. — From $\triangle AOB$ obtain the equation,
$256 + x^2 = (x+8)^2$.

(b) If $BO = 10$ and $EP = 6$, find $AP\ (= x)$.

Ans. $11\frac{1}{4}$ and 0.

Suggestion. — Use the equation,
$$[6 + \sqrt{x^2+36}]^2 = (x+10)^2 - 100.$$

(c) If $AP = a$ and $EP = b$, find $BO\ (= x)$.

Ans. $\dfrac{b^2 + b\sqrt{a^2+b^2}}{a}$.

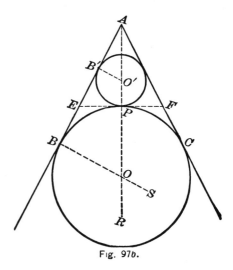

Fig. 97b.

(d) If $OB = a$ and $PE = b$, find $AP\ (= x)$.   Ans. $\dfrac{2\,ab^2}{a^2 - b^2}$.

(e) If $OB = a$ and $AP = b$, find $PE\ (= x)$.   Ans. $\dfrac{ab}{\sqrt{b^2 + 2\,ab}}$.

(f) Make a similar set of equations for circle O'.

3. In Fig. 97 prove that the circular figures $ONFLO$ and $OMELO$ are congruent.

4. If $OE = 6$, find (1) KL; (2) the area bounded by arc MEL and lines OM and OL; (3) the area of the central figure.
Ans. (1) $6(\sqrt{2}+1)$; (2) $9(4-\pi)(3+2\sqrt{2})$; (3) $36(4-\pi)(3+2\sqrt{2})$.

5. If $OE = a$, find (1) KL and (2) the area of the central figure.   Ans. (1) $a(\sqrt{2}+1)$; (2) $a^2(4-\pi)(3+2\sqrt{2})$.

6. If the area of the central figure is s, find OE.

Ans. $(\sqrt{2}-1)\sqrt{\dfrac{s}{4-\pi}}$.

7. If $ML = 24$, find OE and the area of the central figure.
Ans. (1) $12(2-\sqrt{2})$; (2) $288(4-\pi)$.

8. If K is on AB and $AB = 24$, find OE.   Also find OE if $AB = a$.   Ans. (2) $\dfrac{a}{4}(2-\sqrt{2})$.

OVALS FORMED OF 60° ARCS

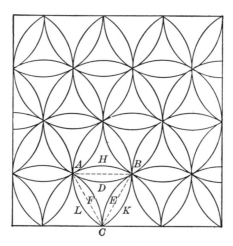

Fig. 98.

EXERCISES

122. **1.** Construct an all-over pattern of equal circles so intersecting that the common chords form equilateral triangles as in Fig. 98.

2. Prove that the ovals formed are congruent.

3. If $AB = a$, find the area of the following curved figures: (a) the oval $AHBD$; (b) the curved $\triangle ADBECFA$.

$$\textit{Ans. } (a) \ \frac{a^2}{6}(2\pi - 3\sqrt{3}); \ (b) \ \frac{a^2}{2}(2\sqrt{3} - \pi).$$

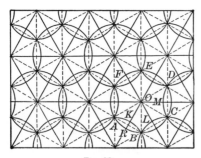

Fig. 99.

Fig. 99a. All-over Embroidery.

EXERCISES

123. 1. Construct an all-over pattern of equal circles so intersecting that the common chords form regular hexagons as in Fig. 99.

2. Prove that the ovals formed are equal and that the circular hexagons are equal.

3. If $AO = a$, find the area of the following curved figures : (a) the oval; (b) the hexagon. *Ans.* (a) $\dfrac{a^2}{6}(2\pi - 3\sqrt{3})$; (b) $a^2(3\sqrt{3} - \pi)$.

CURVED POLYGONS FORMED OF 60° ARCS

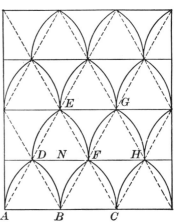

124. Fig. 100 is based on a network of equilateral triangles. The vertices are the centers, and the sides the radii for the arcs.

Fig. 100. — Iron Grill; Stained glass, from Cathedral of Chartres. Shaw (2) Pl. 25.

EXERCISES

1. Prove that the curved figures $BDEF$ and $CFGH$ are congruent.

2. If $AB = 4$, find the area of the following figures: (a) the segment inclosed by the arc AD and the chord AD; (b) $DBFND$; (c) $DNFED$.

Ans. (a) $\frac{4}{3}(2\pi - 3\sqrt{3})$; (b) $\frac{4}{3}(9\sqrt{3} - 4\pi)$; (c) $\frac{4}{3}(4\pi - 3\sqrt{3})$.

3. If $AB = a$, find the areas mentioned in Ex. 2. Also find the area and perimeter of the curved figure $DBFED$.

Ans. (a) $\dfrac{a^2}{12}(2\pi - 3\sqrt{3})$; (b) $\dfrac{a^2}{12}(9\sqrt{3} - 4\pi)$; (c) $\dfrac{a^2}{12}(4\pi - 3\sqrt{3})$;

(d) $\dfrac{a^2}{2}\sqrt{3}$; (e) $\dfrac{4}{3}\pi a$.

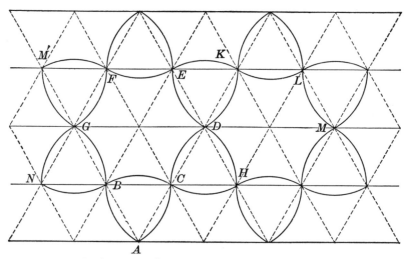

Fig. 101. — From Canterbury Cathedral.   Parker (I), p. 161.

125.   Figure 101 is based on a network of equilateral triangles.   The vertices of the triangles are the centers and the line AB the radius for the arcs drawn as shown.

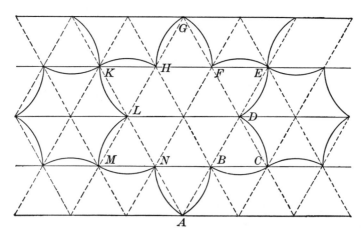

Fig. 102. — Arabic Pattern.   Bourgoin (I), Pl. I.

126.   The construction for Fig. 102 is similar to that for Fig. 101.

EXERCISES

1. In Fig. 101, prove that the curved triangles *ABC, CHD,* etc. are congruent. Also *BCDE* etc. and *DHML* etc.

2. If $AB = 6$, find the area of the following curved figures: (a) *ABC*; (b) *BCDEFG*.

$Ans.$ (a) $18(\pi - \sqrt{3})$; (b) $36(3\sqrt{3} - \pi)$.

3. Find these same areas if $AB = a$.

$Ans.$ (a) $\dfrac{a^2}{2}(\pi - \sqrt{3})$; (b) $a^2(3\sqrt{3} - \pi)$.

4. In Fig. 102, if *AB* is 6 inches, find the perimeter and area of the curved polygon *ABCDEF* etc.

5. If $AB = a$, find the perimeter and area of the figure mentioned in Ex. 4. $Ans.$ (1) $4\pi a$; (2) $2\pi a^2$.

6. In Fig. 101, find the ratio of the area of the curved triangle *ABC* to that of the curved hexagon *ABCDEFG*.

7. Find the ratio of the area of the curved hexagon *ABCDEFG* in Fig. 101 to that of the curved figure *BCD* etc. in Fig. 102, if *AB* in Fig. 101 equals *AB* in Fig. 102.

127. Figure 103 is based on a network of squares. The vertices of the squares are the centers and AB is the radius for the arcs drawn as indicated.

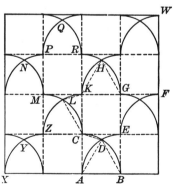

Fig. I03. — All-over Pattern, probably Arabic.

EXERCISES

1. How many degrees in the arcs *LK* and *KH?*

2. Prove that the following curved figures are congruent:

(*a*) *ADC* and *CKL*.

Suggestion. — First prove that the linear triangles *ADC* and *CKL* are congruent. Then superpose the squares *ABEC* and *CKMZ*.

(*b*) *ADB* and *KHG*.

(*c*) *CDEFGHKL* and *LKHRQPNM*.

3. If $AB = 6$, find the area of the curved figure *ABD*. Find this area if $AB = a$. *Ans.* (2) $\dfrac{a^2}{12}(4\pi - 3\sqrt{3})$.

4. If $AB = 6$, find the area of the curved figure *ADECA*. Find this area if $AB = a$. *Ans.* (2) $\dfrac{a^2}{4}(4 - \pi)$.

5. If *AB* is 6 inches, find the area of the curved figure *ADC*. Find this area if $AB = a$. *Ans.* (2) $\dfrac{a^2}{12}(3\sqrt{3} - \pi)$.

6. If *AB* is 6 inches, find the area of the curved figure *CDE*. Find this area if $AB = a$. *Ans.* (2) $\dfrac{a^2}{12}(12 - 2\pi - 3\sqrt{3})$.

7. If $AB = a$, find the area of the curved figure *CDEFG* etc. *Ans.* $2\,a^2$.

8. Draw the figure *CDEFG* etc. on tracing paper, cut along the lines *CE*, *EG*, *GK*, and *KC*. Then rearrange the parts so as to form two complete squares.

9. If $AB = a$, find the lengths of the arcs *AD* and *DE*. *Ans.* $\dfrac{\pi a}{3}$ and $\dfrac{\pi a}{6}$.

10. If $AB = a$, find the perimeter of the figure *CDEFGH* etc. *Ans.* $2\,\pi a$.

11. In passing from point *X* to point *W*, what is the difference between a straight line path and one formed by the arcs *XY*, *YZ*, etc.? *Ans.* $2\,a\,(\pi - 2\sqrt{2})$.

PART 4. DESIGNS FORMED FROM LINE-SEGMENTS AND CIRCLE-ARCS.

128. The designs shown in this part illustrate the possible growth of geometric patterns from very simple units and the possible varieties that may be developed in various industries.[1]

FORMS BASED ON SQUARES ONLY

129. Fig. 104 is the basis for the designs shown in Figs. 105 and 106. In Fig. 104 ABCD is a square with the diameters and diagonals drawn. **XK = XL = XM = YP, etc.**

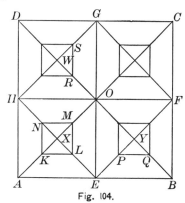

Fig. 104.

EXERCISES

1. Prove that

(a) *AEOH* and *KLMN* are squares.

(b) *K, L, P, Q* are collinear.

(c) *KL* is parallel to *AE*.

2. If $KX = \dfrac{AX}{2}$ and $AB = a$, find the areas of (1) square *KLMN*; (2) trapezoid *AELK*.   *Ans.* (1) $\dfrac{a^2}{16}$; (2) $\dfrac{3\,a^2}{64}$.

3. If $KX = \dfrac{AX}{3}$ and $AB = a$, find the areas mentioned in Ex. 2.

     Ans. (1) $\dfrac{a^2}{36}$; (2) $\dfrac{a^2}{18}$.

4. If $XK = \dfrac{AX}{n}$, find the areas mentioned in Ex. 2.

     Ans. (1) $\dfrac{a^2}{4\,n^2}$; (2) $\dfrac{a^2}{16\,n^2}(n^2 - 1)$.

[1] John Leighton, pp. 119, 120.

5. If $OK = OE$ and if $AB = a$, find the areas mentioned in **Ex. 2.**

Ans. (1) $\dfrac{a^2}{4}(3 - 2\sqrt{2})$;   (2) $\dfrac{a^2}{8}(\sqrt{2} - 1)$.

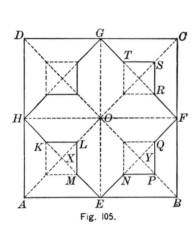

Fig. 105.

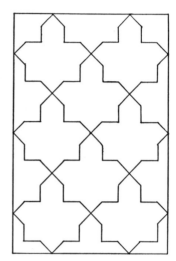

Fig. 105*a*.

EXERCISES

130. **1.** Show that Fig. 105 is derived from Fig. 104.

2. If $AB = 6$ and $XL = \tfrac{1}{2} XO$, find the area of (1) $OHKLMEO$; (2) $OENPQFO$; (3) $ENPQFRS$ etc.

Ans. (1) $3\tfrac{3}{8}$;   (2) $5\tfrac{5}{8}$;   (3) 18.

3. If $AB = a$ and $XL = \tfrac{1}{2} XO$, find the above areas.

Ans. (1) $\dfrac{3\,a^2}{32}$;   (2) $\dfrac{5\,a^2}{32}$;   (3) $\dfrac{a^2}{2}$.

4. If $AB = a$ and $XL = \dfrac{OX}{n}$, find the above areas.

Ans. (1) $\dfrac{a^2}{8\,n^2}(n^2 - 1)$;   (2) $\dfrac{a^2}{8\,n^2}(n^2 + 1)$;   (3) $\dfrac{a^2}{2}$.

Note. It appears that the area of the figure $ENPQFRST$ etc. is $\tfrac{1}{2}$ the area of the square $ABCD$ whatever be the position of the point P.

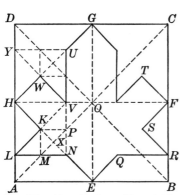

Fig. 106. — Unit for figure 110a. A tile of this shape is on the market.

Fig. 106a. — From the Alhambra. Jones (3), Vol. II, Pl. 9.

EXERCISES

131. 1. Show that Fig. 106 is based on Fig. 104.

2. If $AM = 2\,MX$ and NM is extended to meet AD at L, prove that LK is parallel to AX.

Suggestion. — Since $NE = \frac{1}{3}\,HE$ and MN is parallel to AE, $LA = \frac{1}{3}\,HA$. Since $LA = \frac{1}{3}\,HA$ and $KX = \frac{1}{3}\,HX$, LK is parallel to AX.

3. Extend LK to meet HO at V. Prove that the points N, P, and V are collinear.

Suggestion. — Extend NP to meet HO at V. Join K and V. Prove KV parallel to XO. Therefore LKV is a straight line and coincides with LK extended.

4. Prove that the $\triangle LNK$, the $\triangle NVK$, and the square $MNPK$ are equal in area.

5. Prove that $KVWH$ is a square with vertex V on HO.

6. If $AB = 3$, find the area of (1) $\triangle LKH$; (2) $AENL$; (3) $HKLNEO$; (4) the entire figure $GUVWHKLN$ etc.

7. If $AB = a$, find the above areas.

$$\textit{Ans.} \;\; (1)\;\; \frac{a^2}{36}\,; \;\; (2)\;\; \frac{5\,a^2}{72}\,; \;\; (3)\;\; \frac{11\,a^2}{72}\,; \;\; (4)\;\; \frac{a^2}{2}\,.$$

FORMS BASED ON SQUARES AND SEMICIRCLES

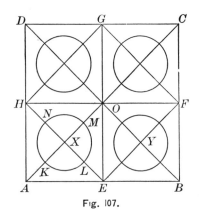

132. In Fig. 107 ABCD is a square with AC and DB the diagonals and EG and FH the diameters. The circles are constructed as indicated.

Fig. 107.

EXERCISES

1. If $KX = \dfrac{AX}{2}$ and $AB = a$, find the area of (1) circle $KLMN$ and (2) figure $AELK$.   *Ans.* (1) $\dfrac{\pi a^2}{32}$; (2) $\dfrac{a^2}{128}(8 - \pi)$.

2. If $KX = \dfrac{AX}{3}$ and $AB = a$, find the above areas.

Ans. (1) $\dfrac{a^2\pi}{72}$; (2) $\dfrac{a^2}{288}(18 - \pi)$.

3. If $KX = \dfrac{AX}{n}$ and $AB = a$, find the above areas.

Ans. (1) $\dfrac{\pi a^2}{8\,n^2}$; (2) $\dfrac{a^2}{32\,n^2}(2\,n^2 - \pi)$.

4. Find the ratio $\dfrac{KX}{AX}$ if the area of the circle $KLMN$ is $\frac{1}{2}$ the area of the square $AEOH$.   *Ans.* $\dfrac{KX}{AX} = \dfrac{1}{\sqrt{\pi}}$.

Suggestion. — Let $KX = \dfrac{AX}{n}$. Solve equation $\dfrac{\pi a^2}{8\,n^2} = \dfrac{a^2}{8}$ for n. Then $KX = \dfrac{AX}{n}$.

5. Find the ratio of $\dfrac{KX}{AX}$ if the area of circle $KLMM$ is $\dfrac{1}{k}$ of the area of the square $AEOH$.   *Ans.* $\dfrac{KX}{AX} = \dfrac{2}{\sqrt{2\,\pi k}}$.

Fig. 108*a*.

Fig. 108*b*. — Counter Railing Design.

EXERCISES

133. 1. Show that Fig. 108 is based on Fig. 107.

2. Find the area of $HKLMEN$ etc, if $AB = a$, (1) if $XM = \dfrac{XE}{2}$, (2) if $XM = \dfrac{XE}{3}$, (3) if $XM = \dfrac{XE}{n}$.

Ans. (1) $\dfrac{a^2}{16}(8 + \pi)$; (2) $\dfrac{a^2}{36}(18 + \pi)$;

(3) $\dfrac{a^2}{4\,n^2}(2\,n^2 + \pi)$.

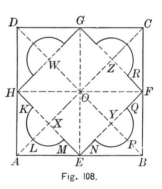

Fig. 108.

3. What must be the ratio $\dfrac{XM}{XE}$ if the semicircle KLM is tangent to the lines HA and AE? *Ans.* $\dfrac{1}{\sqrt{2}}$.

4. Draw a figure to illustrate this case and find the area of $HKLMENPQFR$ etc. as so formed. *Ans.* $\dfrac{a^2}{8}(4 + \pi)$.

Suggestion. — In this case $XM = \dfrac{XE}{\sqrt{2}}$. Substitute $\sqrt{2}$ for n in the result obtained in Ex. 2, part 3.

5. Find the ratio $\dfrac{XM}{XE}$ if the area of $HKLMENPQFR$ is $\frac{3}{4}$ of the area of the square.   *Ans.* $\dfrac{XM}{XE} = \dfrac{1}{\sqrt{\pi}}$.

Suggestion. — Let $XM = \dfrac{XE}{n}$. Solve the equation $\dfrac{a^2(2\,n^2 + \pi)}{4\,n^2} = \dfrac{3\,a^2}{4}$ for n. Then $\dfrac{XM}{XE} = \dfrac{1}{n}$.

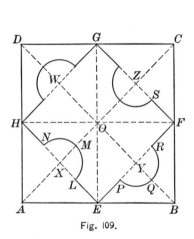

Fig. 109.

Fig. 109a.

EXERCISES

134. 1. Show that Figs. 109 and 109a are based on Fig. 107.

2. If $AB = a$ and $XL = \frac{1}{2}\,XE$, find the area of (1) $HAELMN$ and (2) $HNMLEOH$.   *Ans.* (1) $\dfrac{a^2}{64}(8 + \pi)$; (2) $\dfrac{a^2}{64}(8 - \pi)$.

3. If $AB = a$ and $XL = \frac{1}{3}\,XE$, find the above areas.

Ans. (1) $\dfrac{a^2}{144}(18 + \pi)$; (2) $\dfrac{a^2}{144}(18 - \pi)$.

4. If $AB = a$ and $XL = \dfrac{XE}{n}$, find the above areas.

Ans. (1) $\dfrac{a^2}{16\,n^2}(2\,n^2 + \pi)$; (2) $\dfrac{a^2}{16\,n^2}(2\,n^2 - \pi)$.

5. Show that the area of the figure $HNMLEPQ$ etc. is $\frac{1}{2}$ the area of the square $ABCD$ regardless of the length of the radius XL.

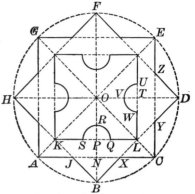

Fig. 110. — Linoleum Design.

135. In Fig. 110 the circle is divided into eight equal parts and the points A, C, E, G, and B, D, F, H are joined. Figure KSRQLW etc. is derived from BDFH as in Fig. 107.

EXERCISES

1. Prove that $ACEG$ and $BDFH$ are congruent squares, and that $BX = XC = CY = YD$, etc.

2. If $OH = a$, find the area of $KSRQLW$ etc. if (1) $PS = \frac{1}{3} PK$ and (2) $PS = \frac{PK}{n}$.   Ans. (1) $\frac{a^2}{18}(18 - \pi)$; (2) $\frac{a^2}{2\,n^2}(2\,n^2 - \pi)$.

3. If $HB = a$, find this area if (1) $PS = \frac{1}{3} KP$ and (2) $PS = \frac{KP}{n}$.

Ans. (1) $\frac{a^2}{36}(18 - \pi)$; (2) $\frac{a^2}{4\,n^2}(2\,n^2 - \pi)$.

4. Compare the areas of $\triangle\,OPL$ and ONC by using the theorem: " The areas of two similar triangles are to each other as the squares on two corresponding sides."

5. If $AC = a$, find the area of trapezoid $ACLK$.   Ans. $\frac{a^2}{8}$.

Suggestion. — Show that the trapezoid is $\frac{1}{2} \triangle AOC$ and hence $\frac{1}{8}$ of square $ACEG$.

6. If $AC = a$ and $PS = \frac{KP}{n}$, find the area of $AJBXCLQRSK$.
Ans. $\frac{a^2}{16\,n^2}(14\,n^2 - 8\,n^2\sqrt{2} + \pi)$.

7. If $AO = r$ and $PS = \frac{KP}{n}$, find the areas of (1) $KSRQLWVU$ etc. and (2) $AJBXCLQRSK$.
Ans. (1) $\frac{r^2}{2\,n^2}(2\,n^2 - \pi)$; (2) $\frac{r^2}{8\,n^2}(14\,n^2 - 8\,n^2\sqrt{2} + \pi)$.

DESIGNS RELATED TO THE PRECEDING

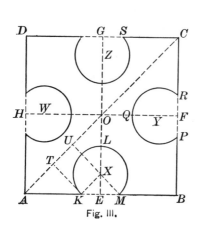

Fig. III.

Fig. IIIa. — Linoleum Design.

EXERCISES

136. 1. In Fig. 111 $ABCD$ is a square with diameters EG and HF. Show how to construct the arc MLK so that $KE = \frac{1}{4} AE$ and $\angle KXM$ is 90°.

2. If $AB = a$, find the areas of (1) segment MLK, and (2) the figure $AKLMBP$ etc., if $\angle KXM$ is 90° and if $KE = \frac{1}{4} AE$.

$$Ans. \ (1) \ \frac{a^2}{128}(2 + 3\,\pi); \ (2) \ \frac{3\,a^2}{32}(10 - \pi).$$

3. If $AB = a$, find the areas as in Ex. 2, except that $KE = \dfrac{AE}{n}$.

$$Ans. \ (1) \ \frac{a^2}{8\,n^2}(2 + 3\,\pi); \ (2) \ \frac{a^2}{2\,n^2}(2\,n^2 - 2 - 3\,\pi).$$

4. What must be the ratio $\dfrac{KE}{AE}$ if circles W and X are tangent to each other and if $\angle KMX$ is 90°?   *Ans.* $\dfrac{KE}{AE} = \frac{1}{2}$.

Suggestion. — If KT and XU are both perpendicular to AO, $AT = TK$ and $OU = UX$. If $KX = XU$, then $AT = TU = UO$. Therefore $AK = KE$.

5. Construct such a figure in which KE shall be $\frac{1}{6} AE$ and $\angle KXM$ shall be 60°.

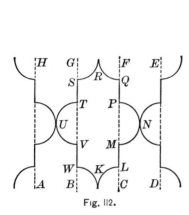

Fig. 112.

Fig. 112a. — Wicker Design.

137. 1. If the distance between the parallels *AH*, *GB*, etc. in Fig. 112 is given, show how to construct the figure.

2. If the distance between any two parallels is 4 inches, find the area of the figure *KLMNPQRST* etc.

Ans. 48 sq. in.

3. If the distance between any two parallels is *a*, find the area of the figure *KLMNPQRST* etc. *Ans.* 3 a^2.

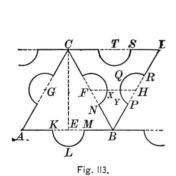

Fig. 113.

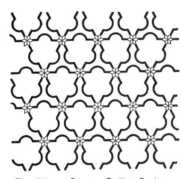

Fig. 113a. — Counter Railing Design.

138. In Fig. 113 △ ABC and BDC are equilateral. E, F, G, H, etc., are the middle points of the sides AB, BC, AC, BD, etc. The semicircles whose centers are E, F, G, H, etc. are drawn with equal radii.

EXERCISES

1. If $KE = \frac{1}{3} AE$ and $AB = 2$, find the area of (1) $AKLMBN$ etc. and (2) $BPQRDS$ etc.   *Ans.* (1) $\left(\sqrt{3} + \frac{\pi}{6}\right)$; (2) $\left(\sqrt{3} - \frac{\pi}{6}\right)$.

2. If $KE = \frac{1}{3} AE$ and $AB = a$, find the same areas.

Ans. (1) $\frac{a^2}{24}(6\sqrt{3} + \pi)$; (2) $\frac{a^2}{24}(6\sqrt{3} - \pi)$.

3. If $KE = \dfrac{AE}{n}$ and $AB = a$, find the same areas.

Ans. (1) $\frac{a^2}{8\,n^2}(2\,n^2\sqrt{3} + 3\,\pi)$; (2) $\frac{a^2}{8\,n^2}(2\,n^2\sqrt{3} - 3\,\pi)$.

4. What must be the ratio of FN to FB in order that semicircles with centers F and H be tangent?   *Ans.* $\dfrac{FM}{FB} = \dfrac{1}{2}$.

5. If $FN = \frac{1}{3} FB$, what is the distance between semicircles with centers F and H if $AB = a$?   *Ans.* $\dfrac{a}{6}$.

6. If $FN = \dfrac{FB}{n}$, find the distance between semicircles with centers F and H if $AB = a$.   *Ans.* $\dfrac{a}{2\,n}(n - 2)$.

7. Construct a figure in which the three semicircles with centers F, H, and T are tangent to each other. If $AB = a$, find the area of the triangular space inclosed by the three semicircles.

Ans. $\dfrac{a^2}{32}(2\sqrt{3} - \pi)$.

8. If $AB = a$ and the semicircles with centers F and H are tangent, find the area of space inclosed by lines NB and BP and the two arcs PY and NX.   *Ans.* $\dfrac{a^2}{48}(3\sqrt{3} - \pi)$.

9. Upon $PR = \frac{1}{3} BD$ as a chord construct a segment of a circle which shall be tangent to the altitude BT. See § 253 (1). If three such arcs are constructed in the $\triangle BCD$, will they be tangent to each other?

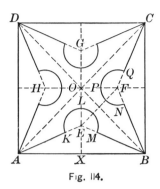

Fig. 114.

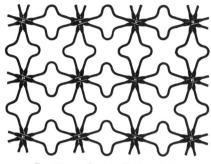

Fig. 114a. — Counter Railing Design.

139. In Fig. 114 ABCD is a square with diameters EG and FH and diagonals AC and BD. KE = EM = FN = FQ, etc., and ∠EAB = ∠EBA = ∠FBC = ∠FCB = ∠GCD, etc.

EXERCISES

1. Prove that (1) AE, EB, and OX are concurrent; (2) $\triangle AEB \cong \triangle FBC$, etc.

Suggestion for (1), OX is the perpendicular bisector of AB; since $AE = EB$, E lies in line OX.

2. If ∠EAB is 30°, KE is $\frac{1}{6} AE$, and $AB = a$, find the areas of (1) $AKLMBA$; (2) $AKLMBNPQC$, etc.

$$Ans. \ (1) \ \frac{a^2}{324} (27\sqrt{3} + 2\pi); \ (2) \ \frac{a^2}{81}(81 - 27\sqrt{3} - 2\pi).$$

3. Find the areas mentioned in Ex. 2 under the same conditions, except that $KE = \dfrac{AE}{n}$.

$$Ans. \ (1) \ \frac{a^2}{36\,n^2}(3\,n^2\sqrt{3} + 8\pi); \ (2) \ \frac{a^2}{9\,n^2}(9n^2 - 3n^2\sqrt{3} - 8\pi).$$

4. If ∠EAB is 30°, construct the figure so that the sectors with centers E and F shall be tangent to each other. Find the radius of each sector in this case. $Ans. \ \dfrac{a}{12}(3\sqrt{2} - \sqrt{6}).$

Suggestion. — $EX = \dfrac{a}{6}\sqrt{3}$; therefore $OE = \dfrac{a}{6}(3 - \sqrt{3})$ and $EF = \dfrac{a}{6}(3\sqrt{2} - \sqrt{6}).$

5. If AE, EB, BF, etc., bisect ⦞ OAB, OBA, OBC, etc., find the length of EX, OE, and EA, if $AB = a$.

Ans. (1) $\frac{a}{2}(\sqrt{2} - 1)$; (2) $\frac{a}{2}(2 - \sqrt{2})$; (3) $\frac{a}{2}(\sqrt{4 - 2\sqrt{2}})$.

Suggestion.—To get EX, use the theorem: "The bisector of an angle divides the opposite side into parts proportional to the other two sides."

6. If AE, EB, BF, etc., bisect ⦞ OAB, OBA, OBC, etc., find the area of (1) $AKLMBA$ and (2) of $AKLMBNP$ etc., if $AB = a$, and $KE = \frac{1}{4} AE$.

$$\text{Ans. (1)} \quad \frac{a^2}{256}(\sqrt{2} - 1)(64 + 5\pi\sqrt{2});$$

$$(2) \quad \frac{a^2}{64}(128 - 64\sqrt{2} - 10\pi + 5\pi\sqrt{2}).$$

7. If AE, EB, BF, etc. bisect ⦞ OAB, OBA, OBC, etc., and $KE = \dfrac{AE}{n}$, find the areas mentioned in Ex. 6.

$$\text{Ans. (1)} \quad \frac{a^2}{16\,n^2}(\sqrt{2} - 1)(4\,n^2 + 5\pi\sqrt{2});$$

$$(2) \quad \frac{a^2}{4\,n^2}(8\,n^2 - 4\,n^2\sqrt{2} - 10\pi + 5\pi\sqrt{2}).$$

PART 5.   DESIGNS BASED ON REGULAR POLYGONS

DESIGNS BASED ON THE REGULAR HEXAGON

140. Occurrence. — All the figures in this group contain the star formed by joining every second vertex of the regular hexagon. It is extremely common in Arabic and medieval[1] ornament and in modern designs. Over the bishop's throne in the cathedral at Anagni is a design executed in mosaic and based on this star,[2] that is worthy of mention. It seems to be an ancient symbol of Deity,[3] and is said to be a Jewish talisman. It is in use also in such modern instances as the policeman's star and many common trademarks.

[1] Wyatt, Pl. IX, Bourgoin (1), Plates.   [2] Arch. Record, XII, p. 217.
[3] Waring (2), p. 66.

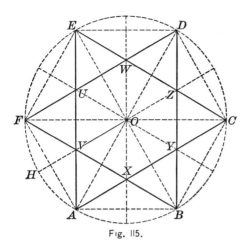

Fig. 115.

Fig. 115*a*. — Border,
Parquet Flooring.

141. In **Fig. 115** the circumference of the circle is divided into **6 equal parts and alternate division points joined.**

EXERCISES

1. Prove that two equilateral triangles are formed.

2. Prove that points V, O, Z, and A, O, D are collinear.

Suggestion. — Draw radius OH to the center of the arc FA. Prove that OH is the perpendicular bisector of FA. Prove $FV = AV$. Therefore V lies on OH.

3. Prove that $\triangle AXV$, BXY, CYZ, etc. are equilateral and congruent.

4. Prove that $XYZW$ etc. is a regular hexagon.

5. Prove that BF is the perpendicular bisector of AO.

6. If $AO = r$, find the length of BF. *Ans.* $r\sqrt{3}$.

7. If $AB = r$, find the area of $\triangle AVX$. *Ans.* $\dfrac{r^2}{12}\sqrt{3}$.

8. If $AO = r$, find the area of (1) the hexagon $XYZW$ etc.; (2) $\triangle AEC$; (3) star $AXBYC$, etc.

Ans. (1) $\dfrac{r^2}{2}\sqrt{3}$; (2) $\dfrac{3\,r^2}{4}\sqrt{3}$; (3) $r^2\sqrt{3}$.

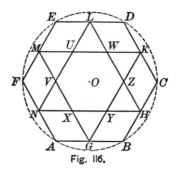

Fig. 116.

Fig. 116a. — T.led Flooring.   Scale ¾ in. l ft.
Medieval, Wyatt, Pl. 3.

142. In Fig. 116 ABCD etc., is a regular hexagon, G, H, K, etc., are the middle points of the sides, and the points are joined as shown in the figure.

EXERCISES

1. Prove that GKM and NHL are congruent equilateral triangles with sides parallel to the sides of $ABCD$ etc.

2. Show that star $GYHZK$ etc. is similar to the star shown in Fig. 115.

3. Prove that $GBHY$ and $HCKZ$ are congruent rhombuses.   Find the size of the angles in these rhombuses.

4. If $AO = r$, find the area of (1) rhombus $BHYG$; (2) $\triangle XGY$; (3) hexagon $XYZW$ etc.; (4) star $GYHZK$ etc.

$$Ans. \; (1) \; \frac{r^2}{8}\sqrt{3}; \; (2) \; \frac{r^2}{16}\sqrt{3}; \; (3) \; \frac{3\,r^2}{8}\sqrt{3}; \; (4) \; \frac{3\,r^2}{4}\sqrt{3}.$$

143. In Fig. 117 ABCDEF is a regular hexagon with each side divided into three equal parts and the points joined as shown.

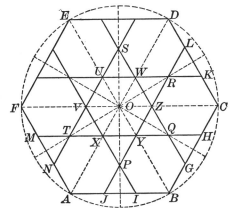

Fig. 117. — Register Design. Arabic. Prisse d'Avenne, Vol. 2, Pl. 103.

EXERCISES

1. Prove that AD, JL, and BC are parallel.

Suggestion. — Extend AB and DC until they intersect. Use the theorem: " A line dividing two sides of a triangle proportionally is parallel to the third side."

2. Prove that points P, O, and S are collinear; also points A, X, O, W, and D; also points N, P, and G.

3. Prove that $BHYJ$, $CLZG$, etc., are congruent rhombuses.

4. Prove that JIP, XPY, GHQ, YQZ, etc., are congruent equilateral triangles.

5. Prove that $IBGQYP$, $CKRZQH$, $XYZWUV$, etc., are congruent regular hexagons.

6. Prove that star $XPYQZR$ etc., is similar to the star shown in Fig. 115.

7. If $AO = r$, find the area of (1) each small hexagon; (2) each small equilateral $\triangle$; (3) the rhombus $BJYH$; (4) the star $XPYQZ$ etc. *Ans.* (1) $\dfrac{r^2}{6}\sqrt{3}$; (2) $\dfrac{r^2}{36}\sqrt{3}$; (3) $\dfrac{2\,r^2}{9}\sqrt{3}$; (4) $\dfrac{r^2}{3}\sqrt{3}$.

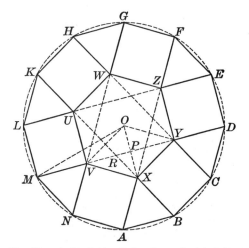

Fig. 118. — Arabic Design Unit. Bourgoin (2), Vol. 7,
II, Pls. 42 and 55.

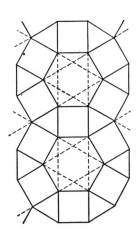

Fig. 118a. — Linoleum
Design.

144. In Fig. 118 the circumference is divided into 12 equal parts and the points joined as indicated.

EXERCISES

1. Prove that

(*a*) ABX, CDY, etc., are congruent equilateral triangles.

(*b*) $BCXY$, $EDYZ$, etc., are congruent squares.

(*c*) $XYZWUV$ is a regular hexagon.

2. Prove that a circle can be passed through the four points A, X, Y, and D.

3. Can a circle be passed through A, Y, W, L?

Find other sets of four points all lying on the same circle.

4. If $OM = r$, find (1) the length of AB; (2) the area of $\triangle ABX$; (3) the area of hexagon $XYZW$ etc.

Ans. (1) $r\sqrt{2 - \sqrt{3}}$; (2) $\dfrac{r^2}{4}(2\sqrt{3} - 3)$; (3) $\dfrac{3r^2}{2}(2\sqrt{3} - 3)$.

Suggestion. — Note that $OP = \frac{1}{2} OV$ and $OV = VM$.

5. If $AB = a$, find (1) length of MO; (2) length of RX; (3) area of the 6-pointed star in the hexagon $XYZW$ etc.

Ans. (1) $a\sqrt{2 + \sqrt{3}}$; (2) $\dfrac{a}{3}\sqrt{3}$; (3) $a^2\sqrt{3}$.

6. Show that Fig. 118a is derived from Fig. 118. Draw to scale [$\frac{3}{4}$ in. = 1 ft.] a diagram for Fig. 118a. Let AB represent 2 inches.

7. Show that Fig. 118a can be constructed from three sets of parallel lines.

8. If $AB = a$, find the distance between the parallels MD and AB in Fig. 118. $\qquad$ *Ans.* $\frac{a}{2}(\sqrt{3}+1)$.

STAR POLYGONS [1]

145. Construction of Star Polygons. — If a circumference is divided into n equal parts and each division point is joined to the kth one from it, where k is an integer greater than one and less than $\frac{n}{2}$, a star polygon is formed. In the discussion which follows these polygons are named from the size of the arcs cut off by the chords. Thus, if the circumference is divided into 8 equal parts and every third point joined, the arc cut off is $\frac{3}{8}$ of the circumference and the star formed is called the $\frac{3}{8}$ star. If a 4-pointed star is derived from the $\frac{3}{8}$ star by omitting part of it, as star *AXCZE* etc., Fig. 121, it is called the 4-pointed $\frac{3}{8}$ star.

146. Occurrence. — The simpler stars are common everywhere in modern ornament and are frequently used in advertising. The occurrence of the two 8-pointed stars is discussed elsewhere. The use of many-pointed stars is largely confined to cut glass. Almost every piece contains one or more. The $\frac{2}{5}$ star is the one used in the United States flag.

147. History. — Star polygons occur in all periods of historic ornament. Their use dates back to the ancient

[1] S. Günther, pp. 1–92.

Greeks when the $\frac{2}{5}$ star was a symbol of recognition among the Pythagoreans.[1] Near Winchester, England, are the remains of an old Roman pavement, the design of which includes the $\frac{5}{16}$ star.[2] In the church of San Giovanni e Paulo at Rome is a Byzantine mosaic which contains the eight-pointed $\frac{7}{16}$ star.[3] Among the mosaics at St. Mark's, Venice, is one that contains the $\frac{2}{5}$ star within the five-pointed $\frac{3}{10}$ star.[4] On the ceiling of one of the halls in the Alhambra is the $\frac{2}{8}$ star within the $\frac{3}{8}$ star.[5] Stars of various kinds are found in medieval tiled pavements in western Europe[6] and in medieval embroideries.[7] They are extremely common in Saracenic art.[8] They occur in the ornamentation of primitive peoples,[9] especially in symbols of the sun.[10]

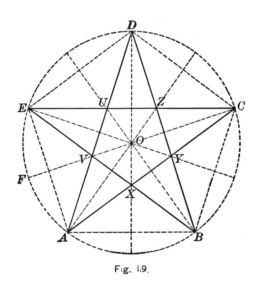

Fig. 119.

[1] Ball, p. 21.   [2] Morgan, p. 221.   [3] Wyatt, Pl. 5.
[4] Ongania, Folio Pl. 30.   [5] Calvert, (1), p. 124.   [6] Shaw, (1), Pl. XXI.
[7] Shaw, (2).   [8] Prisse d'Avennes, Pls. XC, XCI, XCII.
[9] Beauchamp, Pls. 1–5.   [10] Waring, (2), pp. 62 and 63.

148. **1.** Construct the five-pointed star shown in Fig. 119.

2. What axes of symmetry has this figure?

3. Prove that

(*a*) $AC = CE = EB$, etc.

(*b*) $AX = XB = BY$, etc.

(*c*) $\triangle AUC \cong \triangle BVD \cong \triangle CXE$, etc.

(*d*) $\angle BAX = \frac{1}{3} \angle EAB$.

4. Prove that points V, O, C are collinear.

Suggestion. — Prove that each point lies on the perpendicular bisector of AE.

5. Prove that AC is divided into mean and extreme ratio at Y and that AY is divided into mean and extreme ratio at X.

6. Prove that $\angle YXB = 2\angle YBX$.

7. Prove that $XYZUV$ is a regular pentagon.

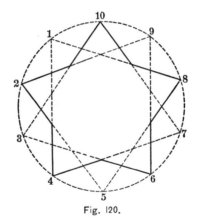

Fig. 120.

149. **1.** Construct the $\frac{3}{10}$ star and the 5-pointed $\frac{3}{10}$ star.

Suggestion. — See Fig. 120.

2. How many 10-pointed stars can be constructed by dividing the circumference into 10 equal parts?

Suggestion. — Every second, third, or fourth point may be joined.

3. Study the composition of each of the 10-pointed stars obtained in Ex. 2.

Suggestion. — From Fig. 120 it is seen that the 10-pointed $\frac{3}{10}$ star can be divided into two 5-pointed $\frac{3}{10}$ stars.

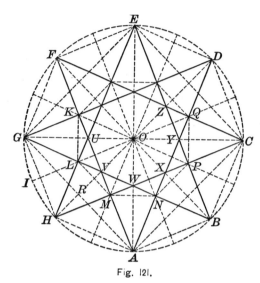

Fig. 121.

Fig. 121*a.* — Parquet Flooring.

EXERCISES

150. **1.** Construct the $\frac{3}{8}$ star in the circle.

Suggestion. — See Fig. 121.

2. What axes of symmetry has the figure?

3. Prove that

(*a*) $AN = NB = BP = PC$, etc.

(*b*) $AB = VA = AX = XC = WB = BY$, etc.

4. Prove that

(*a*) $\angle MAN = \angle NBP$, etc.

(*b*) $\angle HMA = \angle ANB$, etc.

(*c*) $\angle GVA = \angle HWB = \angle AXC$, etc.

(*d*) $\angle MAN + \angle AXC = 2$ rt. $\angle$.

(*e*) $\angle MAN = {}^1 \angle HAB$.

Suggestion. — Turn the figure through an angle of 45.°

5. Find the number of degrees in each angle mentioned in Ex. **4.**

6. Prove that points L, O, and Q, and also H, V, O, Z, and D, are collinear.

7. Prove that $AN = NY$, and that GB is perpendicular to AD.

8. Prove that $KLMNP$ etc., and $UVWXY$ etc. are regular octagons.

9. Prove that the chords GD, HC, AF, and BE intersect in a square.

10. Prove that the figures $ANWM$, $BPXN$, etc. are congruent.
Suggestion. — Turn the figure through an angle of 45°.

11. Prove that the triangles WNX, XPY, YQZ, etc. are congruent.

12. If $AO = a$, find the lengths of HR and HA.
$$Ans.\ HA = a\ \sqrt{2 - \sqrt{2}}.$$

13. If $AO = a$, find the lengths of AM and HC.

• $Ans.$ (1) $\dfrac{a}{2}\sqrt{4 - 2\sqrt{2}}$; (2) $a(\sqrt{2} + 1)(\sqrt{2 - \sqrt{2}}.$

14. If $AO = a$, find the area of square LQ and of $\triangle HMA$.
$$Ans.\ (1)\ a^2(2 - \sqrt{2});\ (2)\ \frac{a^2}{4}(2 - \sqrt{2}).$$

Fig. 121*b*. — Vaulting from Peterborough Cathedral.

15. Construct a diagram for Fig. 121*b*. Are all of the lines straight?

16. Construct the 4-pointed star *A X C Z E* etc. in Fig. 121.

17. If the circumference is divided into 8 equal parts, how many 8-pointed stars are possible?

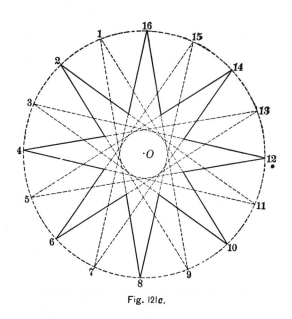

Fig. 121*c*.

18. Construct the $\frac{2}{8}$ star and study its composition.

19. Construct the $\frac{6}{16}$ star and show that it is composed of two $\frac{3}{8}$ stars.

20. If the circumference is divided into 16 equal parts, how many 16-pointed stars can be constructed?

21. Construct the $\frac{2}{16}$ and $\frac{4}{16}$ stars. Study the composition of each.

22. Construct the $\frac{5}{16}$ star. Construct the 4-pointed $\frac{5}{16}$ star and the 8-pointed $\frac{5}{16}$ star.

23. Construct the 4-pointed and the 8-pointed $\frac{7}{16}$ stars.

Suggestion. — See Fig. 121c.

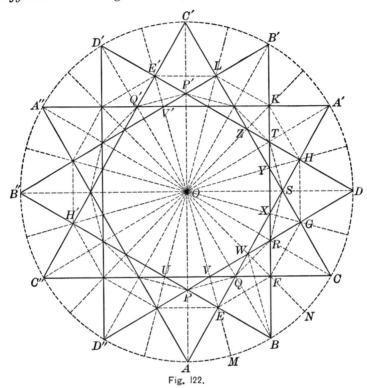

Fig. 122.

EXERCISES

151. **1.** Construct the regular $\frac{4}{12}$ star.

Suggestion. — See Fig. 122.

2. Give the axes of symmetry of this figure.

3. Prove that

(a) $AE = EB = BF = FC$, etc.

(b) $AQ = QC = BR = RD$, etc.

(c) $AW = WD = BX = XA'$, etc.

4. Prove that

(*a*) $\angle A = \angle B = \angle C$, etc.

(*b*) $\angle AEB = \angle BFC = \angle CGD$, etc.

(*c*) $\angle AQC = \angle BRD = \angle CSA'$, etc.

(*d*) $\angle AWD = \angle BXA' = \angle CYB'$, etc.

5. Find the number of degrees in each angle mentioned in Ex. 4.

6. Prove that the points E, V, O, V', and E' are collinear; also the points B, Q, O, Q', and D'.

7. Prove that the figures $BFQE, CGRF$, etc. are congruent. Prove EQF, FRG, etc. congruent isosceles triangles.

8. Prove that $EQVP, FRWQ$, etc. are congruent. Prove that $PV = VQ = QW$, etc.

9. Prove that $OW = WB = AB$, and that $AE = EB = EW$.

10. Prove that (1) $EFGHK$ etc., (2) $PQRST$ etc., (3) $UVWXY$ etc. are regular duodecagons.

Suggestion. — Turn the figure through an angle of 30°.

11. Prove that PRT etc. is a regular hexagon. How many other regular hexagons can be found in the figure?

12. Prove that $EHE'H'$ is a square. How many other squares can be found in the figure?

13. Construct the 4-pointed star $AWDZC'$ etc. How many such stars are there in the figure? Compare it with the 4-pointed $\frac{3}{8}$ star in Fig. 121. Are the two stars congruent when constructed in equal circles?

14. Show that the 12-pointed star in Fig. 122 may be broken up into two $\frac{2}{8}$ stars.

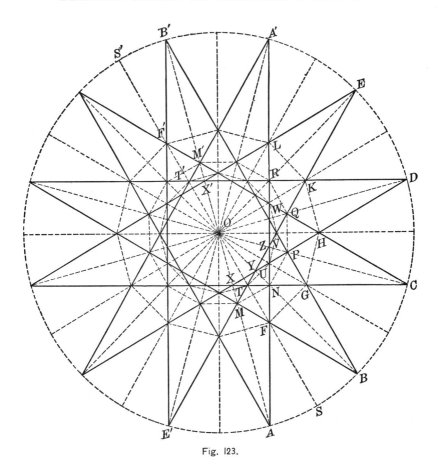

Fig. 123.

152. 1. Construct the $\frac{5}{12}$-star.

Suggestion. — See Fig. 123.

2. Construct the 6-pointed star $ANCQE$ etc. How many such stars are in the figure ?

3. Construct the 4-pointed star AUD etc. How many such stars are in the figure ?

4. Construct the 3-pointed star $AZEX'$ etc. How many such stars are in the figure ?

5. If a circumference is divided into 12 equal parts, how many 12-pointed stars are possible?

6. Explain the composition of the $\frac{2}{12}$ and $\frac{3}{12}$ stars.

7. Construct the $\frac{9}{24}$ star.

8. If a circumference is divided into 24 equal parts, how many 24-pointed stars may be formed?

9. Explain the composition of the $\frac{3}{24}$, $\frac{4}{24}$, $\frac{6}{24}$, and $\frac{8}{24}$ stars.

10. Explain the composition of the $\frac{10}{24}$ star.

11. Construct the $\frac{7}{24}$ and $\frac{11}{24}$ stars. By means of each construct a star of 8 points, one of 4 points, one of 12 points, and one of 6 points.

12. In the figure for the $\frac{3}{8}$ star how many regular octagons occur?

13. In the figure for the $\frac{3}{10}$ star, Fig. 120, how many regular decagons may be formed by joining corresponding intersections?

14. In the figure for the $\frac{4}{12}$ star, how many regular duodecagons occur? See Fig. 122.

15. In the figure for the $\frac{5}{12}$ star, how many regular duodecagons occur? See Fig. 123.

16. In the figure for the $\frac{7}{16}$ star, how many regular polygons of 16 sides occur? See Fig. 121c.

DESIGNS BASED ON STAR-POLYGONS

153. Fig. 124 is the only one of the following designs which is from a modern source. The rest are Mohammedan or Arabic. While only extremely simple Mohammedan designs are here given, they are characteristic of that style, which is highly original, strikingly geometrical, and totally unlike anything else in history. The fact that the Koran forbade the use of all living forms in decoration and art, undoubtedly accounts for its peculiar character. The most complicated interlacing geometrical forms abound in endless variety. It is a style of ornament especially adapted to surface decoration. It is executed in wood, stone, plaster, clay, or mosaic, and completely covers pulpits, domes, vaults, and exterior and interior walls. Window lattices are of the same design. This style characterizes Mohammedan art to-day and has strongly influenced Spanish decoration.

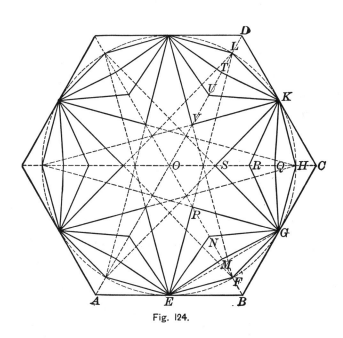

Fig. 124.

154. 1. Construct the design shown in Fig. 124.

Suggestion. — *ABCD* etc. is a regular hexagon with the circle inscribed. *EFGHK* etc. is the regular duodecagon. *EPGSK* etc. is the 6-pointed $\frac{5}{12}$ star. *BF = FM*, *CH = HQ*, etc. *MN = NP*, *QR = RS*, etc.

2. Prove that

(*a*) $FB = CH = DL$, etc.
(*b*) $MF = HQ = LT$, etc.
(*c*) $PN = SR = VU$, etc.
(*d*) $EM = MG = GQ = QK$, etc.
(*e*) $EN = NG = GR = RK$, etc.

3. Prove that

(*a*) Figures $EPGN$, $GSKR$, etc. are congruent.

(*b*) Figures $ENGM$, $GRKQ$, etc. are congruent.

(*c*) Figures $EMGF$, $GHKQ$, etc. are congruent.

Fig. 124*a*. — Parquet Flooring.

(*d*) Figures $EBFG$, $GHKC$, etc. are congruent.

4. Prove that

(*a*) $\angle NGR = \angle RKU$, etc.
(*b*) $\angle MGQ = \angle QKT$, etc.
(*c*) $\angle ENG = \angle GRK$, etc.

5. If $EO = r$, find length of EP. *Ans.* $\dfrac{r}{2}\sqrt{2}$.

6. If $EO = r$, find the area of (1) $\triangle EPG$, (2) star $EPGSK$ etc.

Ans. (1) $\dfrac{r^2}{4}$; (2) $\dfrac{3\,r^2}{2}(\sqrt{3}-1)$.

7. If $EO = r$, find the area of (1) the hexagon $ABCD$ etc., (2) the figure $EPGB$. *Ans.* (1) $2\,r^2\sqrt{3}$; (2) $\dfrac{r^2}{12}(3+\sqrt{3})$.

8. If $EO = r$, find (1) the length of FB and (2) the area of $\triangle EFB$. *Ans.* (1) $\dfrac{r}{3}(2\sqrt{3}-3)$; (2) $\dfrac{r^2}{12}(2\sqrt{3}-3)$.

9. If $EO = r$, find the length of (1) PF, (2) PM, (3) MN.

Ans. (1) $\dfrac{r}{2}(3-\sqrt{3})$; (2) $\dfrac{r}{6}(15-7\sqrt{3})$; (3) $\dfrac{r}{12}(15-7\sqrt{3})$.

10. If $EO = r$, find the area of figure $ENGM$.

Ans. $\dfrac{r^2}{24}(15-7\sqrt{3})$.

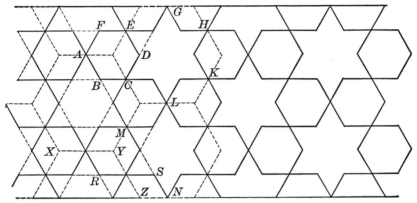

Fig. 125. — Arabic All-over.   Bourgoin (I), Pl. I.

EXERCISES

155. 1. Construct the pattern shown in Fig. 125.

Suggestion. — Construct an all-over pattern of regular hexagons. In each hexagon construct the $\frac{2}{6}$ star shown in Fig. 116.

2. Prove that the points A, M, and N are collinear. Prove that $RS = XY$.

3. Show that Fig. 125 can be constructed from three sets of parallel lines.

4. If $RS = a$, find the perpendicular distance between the parallels and the angle of intersection between the sets of parallels.

$$Ans. \ \frac{a}{2}\sqrt{3}; \quad 60°.$$

5. If $RS = a$, find the area of the hexagon $ABCDEF$.

$$Ans. \ \frac{3\,a^2}{8}\sqrt{3}.$$

6. If $RS = a$, find the area of the star $CLKH$, etc.

$$Ans. \ \frac{3\,r^2}{4}\sqrt{3}.$$

7. Show that Fig. 125 may be constructed from an all-over of equilateral triangles each of whose sides is equal to ZN.

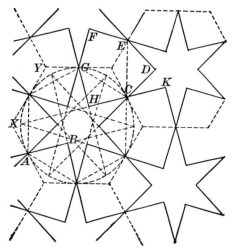

Fig. 126. — Arabic All-over.   Bourgoin (I), Pl. 12.

EXERCISES

156. 1. Construct the pattern shown in Fig. 126.

Suggestion. — Construct an all-over pattern of regular hexagons. In each hexagon inscribe a circle.   In each circle construct the 6-pointed $\frac{5}{12}$ star (see § 152, Ex. 2), so that its points shall be the midpoints of the sides of the hexagon.   Why is this possible?

2. Prove that points H, C, and K are collinear.

3. Prove that lines AB and HK are parallel.

4. Prove that figure $CDEFGH$ is equilateral and that it has two sets of equal angles of three each.   Find the size of each.

5. Prove that all figures of the type $CDEFGH$ are congruent.

6. Prove that if GE is drawn $\triangle\,GEC$ is equilateral.

7. If the radius of one of the circles is r, find the length of GH.

$$Ans.\ \frac{r}{2}\sqrt{2}.$$

8. If the radius of one circle is r, find the area of (1) the star; (2) the figure $CDEFGH$.

$$Ans.\ (1)\ \frac{3\,r^2}{2}(\sqrt{3}-1);\ (2)\ \frac{r^2}{4}(\sqrt{3}+3).$$

9. If a side XY of one of the hexagons is a, find the length of GH.

$Ans. \dfrac{a}{4}\sqrt{6}.$

10. If $XY = a$, find the area of (1) the star; (2) the figure $CDEFGH$. $Ans.$ (1) $\dfrac{9\,a^2}{8}(\sqrt{3} - 1)$; (2) $\dfrac{3\,a^2}{16}(\sqrt{3} + 3).$

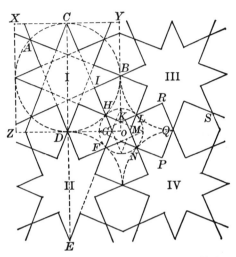

Fig. 127. — Arab c All-over. Bourgoin (I), Pl. 42.

EXERCISES

157. 1. Construct the pattern shown in Fig. 127.

Suggestion. — Construct an all-over pattern of squares. In each square inscribe a circle. In each circle construct the ⅜ star, so that every second vertex will coincide with the middle points of the sides of the square. Construct the small star O by extending lines in stars I, II, III, and IV until they intersect.

2. Prove that points A, B, and S, also C, D, and E, are collinear.

3. Prove that EF, CH, and ZO are concurrent.

4. Prove that

(1) $FG = GH = HK$, etc.

(2) $\angle GHK = \angle KLM$, = etc.

(3) Star $FGHKL$ etc., is the regular 4-pointed ⅜ star.

5. Prove that (1) $LR = RQ = QP = PN$; (2) $ML = MN$.

6. Find the number of degrees in each angle of figure $LMNPQR$.

7. Prove that the figures $LMNPQR$ and $BIHK$ etc. are congruent.

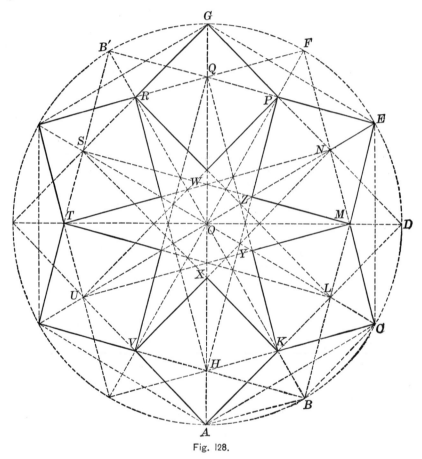

Fig. 128.

EXERCISES

158. **1.** Show how to construct Fig. 128.

Suggestion. — Construct the 6-pointed $\frac{3}{12}$ star $AKCME$ etc. in a given circle. Construct the star $KYMZP$ etc. by joining the proper points.

2. Prove that

(a) $AB = AK = KC = CM$, etc.　　　(c) $\angle AKC = \angle CME$, etc.

(b) $\angle KCM = \angle MEP$, etc.

3. Find the number of degrees in each angle mentioned in Ex. 2.

4. Prove that the points B, K, O, R, and B' are collinear, also the points A, H, O, Q, and G.

5. Prove that $HKLMN$ etc. is a regular duodecagon.

6. Prove that $KCMY$ and $MEPZ$ are congruent squares.

7. Prove that the star whose points are $KLMN$ etc. is a $\frac{5}{12}$ star.

8. Prove that the points O, X, A, and O, Y, C, etc., are collinear.

9. Prove that the circle through points A, K, and C is equal to the given circle.

10. If $OA = r$, find the length of (1) AB; (2) OX; (3) KB.

$Ans.$ (1) $r\sqrt{2 - \sqrt{3}}$; (2) $r(2 - \sqrt{3})$; (3) $r(2 - \sqrt{3})$.

11. If $OA = r$, find the area of (1) the square $CMYK$; (2) the rhombus $ABCK$; (3) the star $AKCME$ etc.; (4) the star $XKYMZ$ etc.　　　$Ans.$ (1) $r^2(2 - \sqrt{3})$; (2) $\frac{r^2}{2}(2 - \sqrt{3})$; (3) $3r^2(\sqrt{3} - 1)$;

(4) $3r^2(3\sqrt{3} - 5)$.

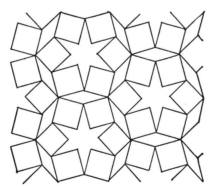

Fig. 128a. — Arabic All-over.　Bourgoin, (I) Pl. 13.

12. Construct the pattern shown in Fig. 128a.

Suggestion. — Construct an all-over pattern of regular hexagons. About each circumscribe a circle.

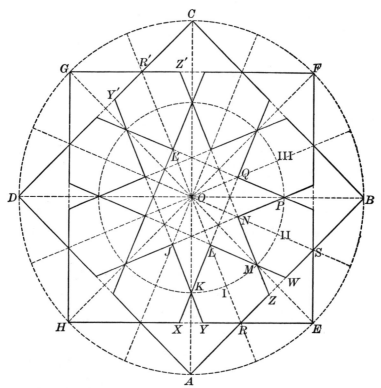

Fig. 129. — Common Unit in Arabic Design.   Jones (3), Vol. II, Pl. XLV.

EXERCISES

159. 1. Show how to construct Fig. 129.

Suggestion. — Construct the $\frac{2}{8}$ star in a given circle.   Let K be any arbitrary point on the radius OA, where A is one vertex of the $\frac{2}{8}$ star. Draw the circle whose center is O and whose radius is OK.   In this circle construct the $\frac{3}{8}$ star with its vertices on the radii OA, OE, OB, etc.   Extend the lines as indicated.

2. Prove that (a) $XK = KY = ZM = MW$, etc.; (b) $HX = YE = AZ = WB$, etc.; (c) $YR = RZ = WS$, etc.

3. Prove that points R, L, O, L', R' are collinear.

4. Find the number of degrees in each angle of figure $RZMLKY$.

5. Prove that figures I, II, and III are congruent.

6. Prove that YY', RR', and ZZ' are parallel.

7. Point K may be determined by bisecting $\angle ORH$. Let the bisector intersect OA at K. Prove that in this case $YK = YR$.

8. Is it possible to make $YR = KL$?

Suggestion. — Is it possible to make $\angle KLR = \angle LRK$?

9. Construct figures similar to Fig. 129, by dividing the circumference into sixteen equal parts, joining every second point, and constructing any sixteen-pointed star desired in the inner circle.

10. Construct a figure like Fig. 129 but of 12 parts.

Suggestion. — See Fig. 129a.

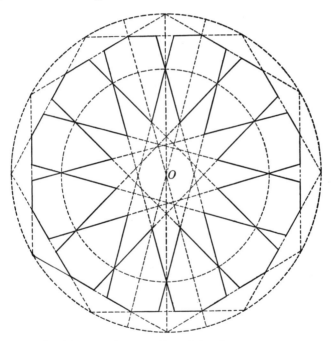

Fig. 129a. — Arabic Design Unit.   Prisse d'Avennes, Pl. 60.

CHAPTER IV

GOTHIC TRACERY: FORMS IN CIRCLES

160. Origin and Development.[1] — The origin of Gothic tracery is evident from a study of English buildings. During the latter part of the twelfth and during the thirteenth century, lancet windows which formerly were used singly were often grouped under a common drip stone. When two lights were so grouped, the stone between them and the arch above was sometimes pierced by a round window. This was the beginning of plate tracery. True bar tracery came later. In it the openings are separated by extensions of the mullions with no solid masses of stone between the tracery bars.

Tracery is usually associated with Gothic windows, but during the middle and later Gothic, the interior and exterior of buildings are profusely decorated with it. It is cut in stone and carved in wood. It adorns altars, screens, seats, pulpits, and all the interior furnishings of churches and cathedrals. The designs, at first very simple, became later most complicated and elaborate.

---

[1] Bond, Chap. XXXI; Fergusson, Vol. II., p. 163; Sturgis (1), p. 268; Sharpe (1).

PART 1. FORMS CONTAINING CIRCLES INSCRIBED IN TRIANGLES

161. With one exception, all of the illustrations used in this section are from **decorated rafters.**

History and Occurrence. — Timber roofs have been in use whenever and wherever lumber was abundant enough to be used for building. In the early Christian and in the Romanesque buildings the use of the circular arch left rectangular spaces between the arch and the rafters or ceilings above, which often received ornament more or less elaborate.

In the Middle Ages highly ornamented **open-timber roofs** abounded in England.[1] The hammer-beam roof of the fifteenth century, especially, afforded opportunity for the use of tracery. (See Fig. 132*a*.)

The modern methods of artificial heating in winter render it impossible to use the main structural features of the roof for ornamental purpose as was done in the fourteenth and fifteenth centuries. The ornamented rafters seen in modern buildings, therefore, are not the rafters that support the roof.

162. In Fig. 130 AFEC is a rectangle circumscribed about the semicircle AFB.

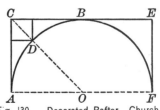

Fig. 130. — Decorated Rafter. Church of San Miniato, Florence. Fletcher, Pl. 93.

[1] Fletcher, p. 290; Brandon (2), Pls. 18, 19, 20.

1. Find the relation between AC and AF.

2. If O is the center of the semicircle, prove that $\angle AOC$ is one half a right angle.

3. Construct the square with CD as diagonal.

4. If $AO = r$, find the area of the square CD.

$$Ans. \ \frac{r^2}{2}(3 - 2\sqrt{2}).$$

5. If $AC = a$, find the area of the figure bounded by the arc ADB and the tangents AC and BC. $Ans. \ \dfrac{a^2}{4}(4 - \pi).$

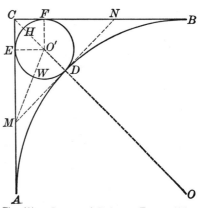

163. In Fig. 131 ACB is a right angle with AC = CB.   Circle ADB is tangent to AC and CB at A and B respectively.   Circle O′ is tangent to AC, CB, and to the arc ADB.

Fig. 131. — Decorated Rafter or Tracery Window Design.

1. Show how to find the center of the circle ADB.

Suggestion. — Erect perpendiculars to AC and CB at A and B, respectively.

2. Prove that $AOBC$ is a square.

3. Is the problem, " To construct a circle tangent to the sides of an angle at given points on these sides," generally possible?   Why is it possible in this case?

4. Show how to find the center of circle O'.

5. Show that the problem given in Ex. 4 is a special case of the problem, " To construct a circle tangent to two given intersecting lines and to a given circle." Why is this special case capable of a simple solution?

6. Show that the point O' divides CD so that $\dfrac{CO'}{O'D} = \dfrac{\sqrt{2}}{1}$.

7. If $CA = a$, find the length of CD, CO', and $O'D$.
   Ans. (1) $a(\sqrt{2} - 1)$; (2) $a(3\sqrt{2} - 4)$; (3) $a(3 - 2\sqrt{2})$.

8. If $AC = a$, find the area of the small circle.
   Ans. $\pi a^2(17 - 12\sqrt{2})$.

9. If $AC = a$, find the area of the figure bounded by $\overparen{EHF}$ and the line-segments CE and CF.   Ans. $\dfrac{a^2}{4}(17 - 12\sqrt{2})(4 - \pi)$.

10. If $AC = a$, find the area of the figure bounded by $\overparen{EWD}$, $\overparen{AD}$, and line-segment AE.   Ans. $\dfrac{a^2}{2}(12\sqrt{2} - 16 - 13\pi + 9\pi\sqrt{2})$.

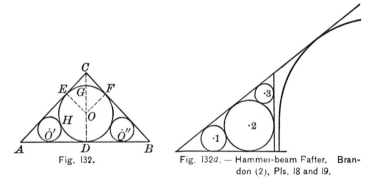

Fig. 132.   Fig. 132a. — Hammer-beam Rafter. Brandon (2), Pls. 18 and 19.

EXERCISES

164. 1. In Fig. 132 if $CA = CB$ and $\triangle ACB$ is a right triangle with right angle at C, construct the circles O, O' and O''.

2. If $AB = a$, find the length of OD.   Ans. $\dfrac{a}{2}(\sqrt{2} - 1)$.

3. If $AB = a$, find the area of circle O.   Ans. $\dfrac{a^2\pi}{4}(3 - 2\sqrt{2})$.

4. If $AB = a$, find the area bounded by $\overparen{EGF}$ and the tangents CE and CF.   Ans. $\dfrac{a^2}{16}(4 - \pi)(3 - 2\sqrt{2})$.

5. If $AB = a$, find the area of the quadrilateral $AEOD$.

$$Ans.\ \frac{a^2}{4}(\sqrt{2} - 1).$$

6. If $AB = a$, find the area bounded by $\overset{\frown}{EHD}$ and the tangents EA and AD. $Ans.\ \dfrac{a^2}{32}(8\sqrt{2} - 8 - 9\pi + 6\pi\sqrt{2}).$

7. If $CA = a$, find the area of circle O. $Ans.\ \dfrac{a^2\pi}{2}(3 - 2\sqrt{2}).$

8. If $AC = a$, find the area of (1) the quadrilateral $AEOD$; (2) the figure bounded by $\overset{\frown}{EGF}$ and the tangents CE and CF; (3) the figure bounded by $\overset{\frown}{EHD}$ and the tangents EA and AD.

$$Ans.\ (1)\ \frac{a^2}{2}(\sqrt{2} - 1);\ (2)\ \frac{a^2}{8}(4 - \pi)(3 - 2\sqrt{2});$$

$$(3)\ \frac{a^2}{16}(8\sqrt{2} - 8 - 9\pi + 6\pi\sqrt{2}).$$

9. Construct circles 1, 2, and 3 in any right triangle as shown in Fig. 132a.

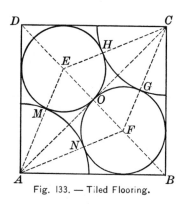

Fig. 133. — Tiled Flooring.

EXERCISES

165. **1.** In Fig. 133 circles E and F are inscribed in the isosceles right triangles ADC and ABC. Show that arcs HG and MN can be drawn tangent to the circles as indicated.

2. Prove that EHC, FGC, etc., are straight lines.

3. If $AB = a$, find (1) the area of circle E; (2) the length of AE; (3) the length of AM.

Ans. (1) $\dfrac{\pi a^2}{2}(3 - 2\sqrt{2})$; (2) $a\sqrt{2 - \sqrt{2}}$; (3) $a\sqrt{2 - \sqrt{2}} - \dfrac{a}{2}(2 - \sqrt{2})$.

4. Make a drawing for a tiled floor composed of tiles like that shown in Fig. 133. The tiles are to be so placed that the corner quadrants of four tiles meet to form a circle. The tiles are six inches on a side. Make the drawing on a scale 4 inches to 1 foot.

PART 2. ROUNDED TREFOILS

166. History and Occurrence. — The names **trefoil** and **quadrifoil** are applied to various three-leaved and four-leaved forms, respectively. Those considered in the two following sections are the rounded forms and are constructed from equal circles which are tangent or which intersect. They abound in Byzantine and Gothic architecture, especially in the cathedrals of the thirteenth and fourteenth centuries[1] where they form small windows in clerestories and gables, and occur as details in the tracery of large windows, in painted glass, in the wood carvings of interior furnishings, in sculptured stone, and in ornamental iron. They are supposed to symbolize the Trinity and the four Gospels, respectively.[2]

Trefoils have been used since the Middle Ages in nearly every variety of ornament. Examples may be found in almost every town. Illustrations of their occurrence in tracery windows may be found in Figs. 208a, 246a, and 256a. Figure 137a shows two different kinds of trefoils. Many more or less complicated designs have been based on the simpler ones here given.[3]

[1] A. Pugin, Illustrations; A. and A. W. Pugin, Illustrations.

[2] Wornum, p. 106.

[3] Billings (3); Day, p. 36; Brandon (2), Pl. 40.

THE TREFOIL FORMED OF TANGENT CIRCLES

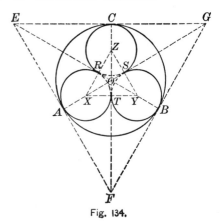

167. Figure 134 shows a trefoil formed of the three circles X, Y, and Z tangent to each other at the points T, S, and R. It is inscribed in the circle as shown.

Fig. 134.

EXERCISES

1. Show how to construct the figure.

Solution. — Circumscribe an equilateral triangle about the circle. Connect each vertex with the center. Inscribe a circle in each of the triangles FOG, GOE, and EFO.

2. Prove that the small circles are tangent to the large circle and to each other.

3. Prove that the curved figures ATB and BSC are congruent.

4. If $OA = r$, find EF.   *Ans.* $2\,r\sqrt{3}$.

5. If $XY = 6$, find OA.   *Ans.* $(2\sqrt{3} + 3)$.

Suggestion. — Find XS, OX, then OA.

6. If $XY = 2\,a$, find OA.   *Ans.* $\dfrac{a}{3}(2\sqrt{3} + 3)$.

7. If $OA = r$, find XT.   *Ans.* $r(2\sqrt{3} - 3)$.

8. If $XY = 6$, find the area of (1) the curved figure RST; (2) the trefoil $TBSCR$ etc.; (3) the curved figure CSB.

Ans. (1) 1.45+; (2) 86.27+; (3) 14.99+.

9. If $XY = 2\,a$, find the areas mentioned in Ex. 8.

Ans. (1) $\dfrac{a^2}{2}(2\sqrt{3} - \pi)$; (2) $\dfrac{a^2}{2}(5\,\pi + 2\sqrt{3})$; (3) $\dfrac{a^2}{18}(8\,\pi\sqrt{3} - 6\sqrt{3} - \pi)$.

10. If $OA = r$, find the areas mentioned in Ex. 8.

$$Ans. \ (1) \ r^2(21\sqrt{3} + 6\,\pi\sqrt{3} - \frac{21\,\pi}{2} - 36);$$

$$(2) \ \frac{3\,r^2}{2}(35\,\pi - 20\,\pi\sqrt{3} + 14\sqrt{3} - 24);$$

$$(3) \ \frac{r^2}{6}(60\,\pi\sqrt{3} - 42\sqrt{3} - 103\,\pi + 72).$$

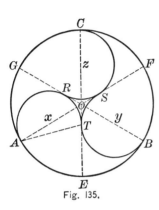

Fig. 135.

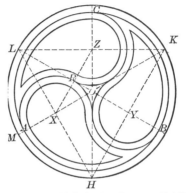

Fig. 135*a*. — Diefenbach. Series B, Pl. VII.

168. **Figure 135 is based on Fig. 134.** Figure 135*a* shows the working drawing for certain parts of Fig. 135*b* and is based on Fig. 135.

Fig. 135*b*. — Scottish Rights Church, St. Louis.

EXERCISES

1. How many degrees in the arc TRA of Fig. 135? *Ans.* 210°.

2. Prove that the curved figures $AEBTR$ and $BFCST$ are congruent.

3. If $AO = r$, find the area of the figure $AEBTR$.

$$Ans. \ \frac{r}{6}(23\,\pi - 12\,\pi\sqrt{3} - 42\sqrt{3} + 72).$$

4. Show how to construct Fig. 135a if the radius of the outer circle is given.

Suggestion. — Points X, Y, and Z are the centers for the small circles. They are determined by the $\triangle$ HKL.

5. If $KM = 2\,a$ in Fig. 135a, find the lengths of (1) HK, (2) XR, (3) OA, and (4) AM.

$$Ans. \ (1) \ a\sqrt{3}; \ (2) \ \frac{a}{4}\sqrt{3}; \ (3) \ \frac{a}{4}(2 + \sqrt{5}); \ (4) \ \frac{a}{4}(2 - \sqrt{3}).$$

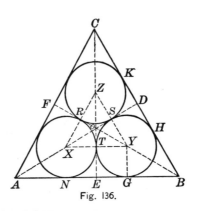

Fig. 136.

EXERCISES

169. 1. Inscribe a trefoil in the equilateral triangle ABC.

Suggestion. — See Fig. 136.   Draw the medians AD, BF, and CE. Inscribe circles in the figures $OEBD$, $ODCF$, etc.

2. If $AB = a$, find the length of GY and OT.

$$Ans. \ (1) \ \frac{a}{4}(\sqrt{3} - 1); \ (2) \ \frac{a}{6}(3 - \sqrt{3}).$$

Suggestion. — Since $BY = 2\,YG$, we have $\dfrac{YG}{GB} = \dfrac{1}{\sqrt{3}}.$

3. If $AB = a$, find the following areas:

(a) The curved figure RST.　　*Ans.* $\dfrac{a^2}{16}(2 - \sqrt{3})(2\sqrt{3} - \pi)$.

(b) The trefoil $TGHSK$ etc.　　*Ans.* $\dfrac{a^2}{16}(2\sqrt{3} + 5\pi)(2 - \sqrt{3})$.

(c) The figure bounded by $\overparen{GH}$ and the line-segments GB and BH.

$$Ans.\ \dfrac{a^2}{24}(2 - \sqrt{3})(3\sqrt{3} - \pi).$$

(d) The figure bounded by $\overparen{NT}$ and $\overparen{TG}$ and the line-segment NG.

$$Ans.\ \dfrac{a^2}{16}(2 - \sqrt{3})(4 - \pi).$$

ROUNDED TREFOILS FORMED OF INTERSECTING CIRCLES

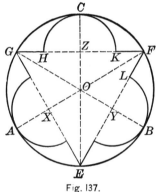

Fig. 137.

Fig. 137a. — From St. Mary's Church, Speenhamland, Newbury, England.

170. In Fig. 137 EFG is an equilateral triangle inscribed in a circle and X, Y, and Z, are the centers of the semicircles constructed as shown.

EXERCISES

1. If $OA = a$, find the lengths of OZ, ZH, and GH.

$$Ans.\ GH = \dfrac{a}{2}(\sqrt{3} - 1).$$

2. Prove that the curved figures GHC and KFC are congruent.

3. Find the ratios (1) $\dfrac{GZ}{GO}$; (2) $\dfrac{GZ}{ZO}$; (3) $\dfrac{HZ}{GO}$.

$$Ans.\ (1)\ \dfrac{\sqrt{3}}{2};\ (2)\ \sqrt{3};\ (3)\ \tfrac{1}{2}.$$

4. If $AO = a$, find the area of the figure bounded by the arcs GC and CH and the line-segment GH.   *Ans.* $\dfrac{a^2}{48}(5\pi - 6\sqrt{3})$.

5. If $AO = a$, find the area of the figure $GHCKFLB$, etc.

Ans. $\dfrac{3a^2}{8}(2\sqrt{3}+\pi)$.

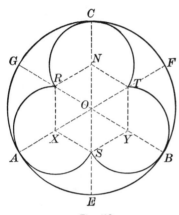

Fig. 138.

Fig. 138a. — From Monk's Church, St. Louis.

171. In Fig. 138 the radii OA, OE, OB, etc., divide the circle into six equal parts, and X, S, Y, T, etc., are the middle points of these radii.

EXERCISES

1. Prove that $XSYT$ etc. is a regular hexagon.

2. If X, Y, and N are the centers of $\overset{\frown}{RAS}$, $\overset{\frown}{SBT}$, and $\overset{\frown}{TCR}$, respectively, prove that these arcs meet at the points R, S, and T.

3. Prove that the curved figures ASB and BTC are congruent.

4. If $AO = a$, find the area of the trefoil $RASBT$ etc.

Ans. $\dfrac{a^2}{8}(3\sqrt{3} + 4\pi)$.

5. Find the area of the curved figure $AEBS$.

6. What fraction of the circle is within the trefoil?

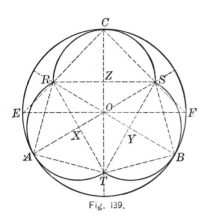

Fig. 139.

Fig. 139a. — From South Park Congregational Church, Chicago.

172. In **Fig. 139** the circumference of the circle is divided into three equal parts by the points A, B, and C. The trefoil ATBSC etc. is composed of semicircles tangent to the circle at points A, B, and C.

EXERCISES

1. Show how to construct the trefoil.

Suggestion. — Construct ∡ OCR and OCS each 45°.

2. Prove that these semicircles intersect on the radii which bisect $\overgroup{AB}$, $\overgroup{BC}$, and $\overgroup{CA}$.

3. Prove that the curved figures ATB and BSC are congruent.

4. If $AO = r$, find the length of AX.   *Ans.* $\dfrac{r}{2}(3 - \sqrt{3})$.

Suggestion. — Show that $\dfrac{OX}{XA} = \dfrac{1}{\sqrt{3}}$.

5. If $AO = r$, find the area of (a) the trefoil $ATBSC$ etc. and (b) the curved figure $ATBA$.

Ans. (a) $\dfrac{3\,r^2}{4}(2 - \sqrt{3})(3\pi + 2\sqrt{3})$; ($b$) $\dfrac{r^2}{12}(18 - 14\pi - 12\sqrt{3} + 9\pi\sqrt{3})$.

PART 3.   ROUNDED QUADRIFOILS

173. Occurrence. — While the quadrifoil is essentially a Gothic form, it is extensively used in **modern industrial ornament.** This may be due to the very general use of the square as the basis for modern design. With the exception of the eight and sixteen foiled figures none of the others are well adapted to such designs.

In **Gothic buildings** they are extremely abundant. The following illustrate their use in tracery windows: Figures 161*a*, 193*a*, 205*a*, 207*a*, 216*a*, 229*a*, 230*a*, 237*a*, 262*a*. When used in stone cutting and wood carving, the two forms are frequently combined in one figure, as in Figs. 149 and 151.

Some special instances of the use of the quadrifoil in **medieval tiled pavements** are noticeable. In the pavement in Chertsey Abbey, England, is one based on Fig. 140. The complete design is a square covering four tiles, each $7\frac{1}{2}$ inches on a side. The circle containing the quadrifoil is surrounded by a handsome scroll border, and all spaces are fully decorated with a fine all-over pattern.[1]

In Gloucester Cathedral and in Great Malvern, Worcestershire, England, are tiled pavements containing designs based on Fig. 142. The first dates back to 1455.[2]

There are a multitude of designs more or less closely related to quadrifoils which are common not only in medieval and modern work, but also in the Chinese and Japanese work and in the decorations of primitive people.[3] Among them is the Chinese Monad used as a charm by the Chinese and as a trademark by the Northern Pacific Railroad (§ 186, Ex. 5, and Remark). (See also Figs. 142 and 152.) The quadrifoil is also found in American Indian decoration.[4]

[1] Shaw (1), Pl. XX.

[2] Shaw (1), Pls. XXXVIII, XXXIX, XLI.

[3] "Wonderland," 1901.   Northern Pacific R. R.   Diefenbach, Series B, Pls. VII, XXIII, XXX.

[4] Beauchamp, Pl. 3.

THE ROUNDED QUADRIFOIL FORMED OF TANGENT CIRCLES

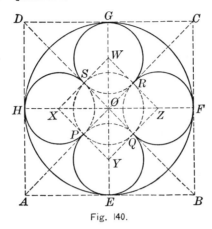

Fig. 140.

EXERCISES

174. **1.** Construct a quadrifoil of tangent circles in a square, so that each circle shall be tangent to one side of the square at its middle point and to two of the other circles.

Suggestion. — See Fig. 140. Draw the diagonals of the square. Inscribe a circle in each triangle thus formed.

2. Prove that the centers of the circles lie on the diameters of the square and that the lines joining their centers form a square.

3. Prove that the curved figures EQF and FRG and also $BEQF$ and $CFRG$ are congruent.

4. Inscribe a quadrifoil of tangent circles in a given circle.

Suggestion. — Circumscribe a square about the circle and inscribe the quadrifoil in the square as in Ex. 1.

5. If $AB = a$, find the length of EY. *Ans.* $\dfrac{a}{2}(\sqrt{2}-1)$.

6. If $AB = a$, find the following areas : (1) the quadrifoil $EQFR$ etc.; (2) the small curved figure $PQRS$; (3) the cusped figure HPE; (4) the cusped figure $HAEP$.

Ans. (1) $\dfrac{a^2}{4}(3-2\sqrt{2})(3\pi+4)$; (2) $\dfrac{a^2}{4}(4-\pi)(3-2\sqrt{2})$;

(3) $\dfrac{a^2}{8}(3\pi\sqrt{2}+4\sqrt{2}-4\pi-6)$; (4) $\dfrac{a^2}{16}(6\pi\sqrt{2}+8\sqrt{2}-9\pi-8)$.

7. If the area of the quadrifoil $EQFRS$ etc. is 144 square inches, find AB.

8. If the area of the quadrifoil is s, find AB.

$$Ans. \quad 2(1 + \sqrt{2}) \sqrt{\frac{s}{3\,\pi + 4}}.$$

9. If the area of the cusped figure HPE is s, find AB.

10. If a circle is inscribed in the square $XYZW$, find the area of the oval PQ and of the cusped figure bounded by $\overset{\frown}{PEQ}$ and $\overset{\frown}{PQ}$. Let $AB = a$.

$$Ans. \quad (1)\ \frac{a^2}{8}\,(3 - 2\,\sqrt{2})\,(\pi - 2);\ (2)\ \frac{a^2}{8}\,(3 - 2\,\sqrt{2})\,(\pi + 2).$$

11. If $XY = b$, find the following areas: (a) the quadrifoil $PEQFR$ etc.; (b) the oval PQ; (c) the cusped figure bounded by $\overset{\frown}{PEQ}$ and $\overset{\frown}{PQ}$.

$$Ans. \quad (a)\ \frac{b^2}{4}\,(3\,\pi + 4);\ (b)\ \frac{b^2}{8}\,(\pi - 2);\ (c)\ \frac{b^2}{8}\,(\pi + 2).$$

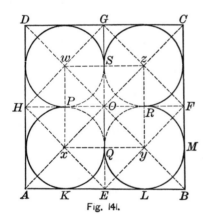

Fig. 141.

Fig. 141a. — Inlaid Tile Design.
Scale $\tfrac{3}{4}$ in. = 1 ft.

EXERCISES

175. **1.** Construct the quadrifoil of tangent circles inscribed in a square so that each circle is tangent to two sides of the square and to two other circles.

Suggestion. — See Fig. 141.

2. Prove that the lines joining the centers of the circles form a square.

3. If $AB = a$, find the following areas :

(*a*) Of the quadrifoil $KQLMR$ etc.   *Ans.* $\dfrac{a^2}{16}(3\pi + 4)$.

(*b*) Of the small curved figure $PQRS$.   *Ans.* $\dfrac{a^2}{16}(4 - \pi)$.

(*c*) Of the figure bounded by the arcs KQ and QL, and the line-segment KL.   *Ans.* $\dfrac{a^2}{32}(4 - \pi)$.

4. If the area of the quadrifoil is s, find AB.

$$Ans.\ 4\sqrt{\dfrac{s}{4 + 3\pi}}.$$

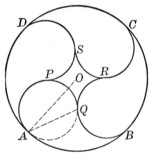

Fig. 142. — From Westminster Abbey Cloisters   Parker ( ), p. 177.

EXERCISES

176. **1.** Show how to construct Fig. 142.

2. How many degrees are there in the arc APQ?   *Ans.* 225°.

3. Prove that the figures $APQB$, $BQRC$, etc., are congruent.

4. If $AO = r$, find the area of the figure $APQB$.

$$Ans.\ \dfrac{r^2}{2}(2\pi - 6 + 4\sqrt{2} - \pi\sqrt{2}).$$

5. If $AO = 2$, find the area of the curved figure $PQRS$.

6. If the area of the figure $PQRS$ is s, find AO.

177. 1. Show how to construct Fig. 143.

Suggestion. — The construction is effected by means of two quadri-foils of tangent circles; w, x, and y are centers for three of the circles.

2. Prove that $xA = OK = OL = OM = ON$, etc.

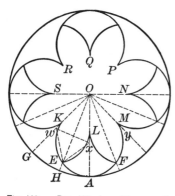

Fig. 143. — Rose Window, Strassengal,
Germany.

Fig. 143a. — Stamped Steel Ceiling
Design.

3. Prove that line OE bisects $\angle GOA$.

Suggestion. — Draw OH the bisector of $\angle GOA$. Prove that $wO = Ox$; OH is the perpendicular bisector of xw; $Ew = Ex$. Therefore, E lies on OH.

4. If $OA = a$, find the area of (1) the cusped figure $EAHG$ etc. and (2) the quadrifoil $KEAFM$ etc.

5. Show that K, L, M, N, etc. are the vertices of a regular octagon.

6. If $OA = a$, find the area of the octagon $KLMN$ etc.

Suggestion. — OK in Fig. 143 is $\frac{1}{2}$ XY in Fig. 144. See also § 88.

7. Draw a diagram for the design shown in Fig. 143a, assuming that all arcs are arcs of circles.

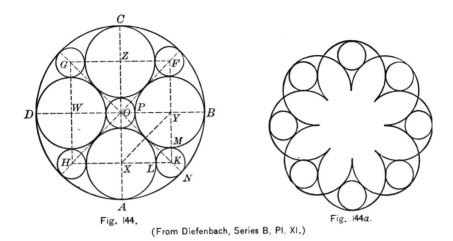

Fig. 144.   Fig. 144a.

(From Diefenbach, Series B, Pl. XI.)

178. In Fig. 144 the circles X, Y, Z, and W are tangent, as in § 174, Ex. 4. The lines HK, KF, FG, and GH are perpendicular to OA, OB, OC, and OD, respectively, at the points X, Y, Z, and W, respectively. The small circles are tangent as indicated.

EXERCISES

1. Prove that *HKFG* is a square and that vertex *K* is on the radius through the point of contact of the circles *X* and *Y*.

2. Prove that (1) $OP = KM = KL$; (2) $OP = KN$.

Suggestion for (2). — $XY = OK = PB$.

3. Prove that circle *K* is tangent to the circles *X* and *Y* and to the large circle *O*.

4. If $OB = a$, find the length of OY, YP, and OP.

    Ans. (1) $a(2 - \sqrt{2})$; (2) $a(\sqrt{2} - 1)$; (3) $a(3 - 2\sqrt{2}.)$

5. If $OP = b$, find OB and PY.

    Ans. (1) $b(3 + 2\sqrt{2})$; (2) $b(\sqrt{2} + 1)$.

6. If $PY = c$, find OB and OP.

    Ans. (1) $c(\sqrt{2} + 1)$; (2) $c(\sqrt{2} - 1)$.

7. What per cent of the area of the large circle is contained within the five small circles *O*, *F*, *G*, *H*, and *K*?   *Ans.* About 15 %.

8. What per cent of the area of the large circle is contained within the four circles X, Y, Z, and W?   *Ans.* 68.62 %.

9. Give the construction for Fig. 144a.

179. In Fig. 145 ABCD is a rectangle with AD = ½ AB.   E and F are the middle points of the sides AB and CD, respectively.   Semicircles are constructed with E and F as centers and quadrants with A, B, C, and D as centers, each with radius = ½ AE.

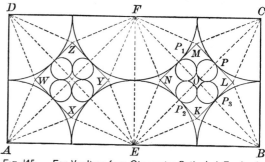

Fig. 145. — Fan Vaulting from Gloucester Cathedral, England. Fletcher, Pl. 112.

EXERCISES

1. Lines are drawn tangent to the arcs at the points where they cut the diagonals of the squares $AEFD$ and $EBCF$. Prove that these tangents form the squares $WXYZ$ and $KLMN$.

2. Show how to construct the entire figure.

3. If $AD = a$, find the radius of one of the small circles.

$$Ans.\ \frac{a}{2}(3 - 2\sqrt{2}).$$

4. If $AD = a$, find the following areas:

(a) The large cusped figure bounded by four quadrants.

$$Ans.\ \frac{a^2}{4}(4 - \pi).$$

(b) The quadrifoil formed by the four small circles.

$$Ans.\ \frac{a^2}{4}(17 - 12\sqrt{2})(3\pi + 4).$$

(c) The small cusped figure included by the four small circles.

$$Ans.\ \frac{a^2}{4}(4 - \pi)(17 - 12\sqrt{2}).$$

(d) The four-sided curved figure bounded by two quadrants and two small circles. $$Ans.\ \frac{a^2}{4}(9\pi\sqrt{2} + 12\sqrt{2} - 13\pi - 16).$$

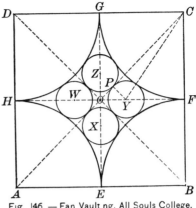

180. In Fig. 146 ABCD is a square, $\widehat{EF}$, $\widehat{FG}$, $\widehat{GH}$, and $\widehat{HE}$ are drawn with A, B, C, and D as centers and the half side AE as radius. The four circles are tangent, as indicated.

Fig. 146. — Fan Vault ng, All Souls College, Oxford. A and A. W. Pugin, Vol. I, Pl. II.

EXERCISES

1. If circle X is tangent either to lines OA and OB, or to arcs HE and EF, show that its center must lie on OE.

2. Construct circle X tangent to the lines OA and OB and to the arcs HE and EF.

Suggestion. — See § 253.

3. Make OY, OZ, and OW each equal to OX. Construct the circles Y, Z, and W each with a radius equal to that of circle X.

4. Prove that circles Y, Z, and W will be tangent to the diagonals of the square and to the arcs as indicated.

5. Prove that each of the four circles will be tangent to two of the others.

6. Find approximately the radius of circle Y if $AB = a$.

Ans. .108 a.

Suggestion. — In $\triangle PYC$, $\overline{PC}^2 + \overline{PY}^2 = \overline{YC}^2$. $PC = \left(\dfrac{a\sqrt{2}}{2} - r\right)$.

$PY = r$, $YC = \left(\dfrac{a}{2} + r\right)$. Get the equation $\left(a\dfrac{\sqrt{2}}{2} - r\right)^2 + r^2 = \left(\dfrac{a}{2} + r\right)^2$,

from which $r = \dfrac{a}{2}(1 + \sqrt{2} \pm \sqrt{2 + 2\sqrt{2}})$.

THE ROUNDED QUADRIFOIL FORMED OF INTERSECTING CIRCLES

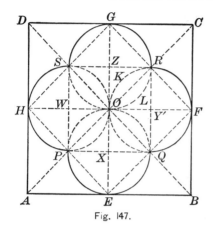

Fig. 147.

Fig. 147*a.* — Oilcloth Design,

EXERCISES

181. 1. Construct in a given square a quadrifoil of equal circles intersecting in a common point, so that the quadrifoil shall be tangent to the sides of the square at their mid-points.

Suggestion. — In Fig. 147, using *OE, OF, OG, OH* as diameters, construct circles.

2. Prove that the circles intersect at the mid-points of the semi-diagonals, *OA, OB, OC, OD,* and that *PQRS* is a square.

3. Prove that arc *PEQ* is a semicircle.

4. Prove that the curved figures *AHPE* and *EQFB* are congruent.

5. Prove that tangents to the two semicircles *PEQ* and *QFR* at their common point *Q* are perpendicular to each other.

6. If *AB = a*, find the area of the following figures: (*a*) the quadrifoil *EQFR*, etc.; (*b*) the oval *OLRK*; (*c*) the cusped figure *AEPH*.

$$\text{Ans.} \quad (a)\ \frac{a^2}{8}(\pi+2);\ (b)\ \frac{a^2}{32}(\pi-2);\ (c)\ \frac{a^2}{32}(6-\pi).$$

7. What per cent of the area of the quadrifoil is occupied by the four ovals?   *Ans.* 22.2 %, about.

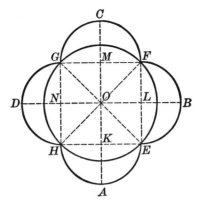

Fig. 148. — Stained-glass Detail.   Holy Trinity Church, Evansville, Indiana.

EXERCISES

182. **1.** Construct Fig. 148.

2. Prove that the crescents AHE and BEF are congruent.

3. If $HE = a$, find the area of (1) the quadrifoil $HAEBF$ etc., and (2) the crescent-shaped figure HAE.

$$Ans. \ (1) \ \frac{a^2}{2}(2 + \pi); \ (2) \ \frac{a^2}{4}.$$

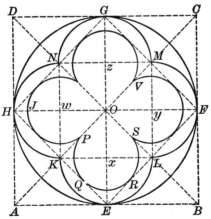

Fig. 149. — From Stone Cutting on the Front of the Church of Our Lady of Good Counsel, New York City.

183. 1. Construct the quadrifoil $ELFMG$ etc. of Fig. 149.

Suggestion. — Use four intersecting circles whose diameters are the radii OE, OF, OG, and OH.

2. Prove that these circles intersect on the radii that bisect $\overarc{EF}$, $\overarc{FG}$, etc.

3. Construct the quadrifoil $PQRS$ etc. in the square $EFGH$.

4. Prove that the points K, P, V, and M are collinear.

5. If $OE = r$, find (1) the radius of circle $PQRS$; (2) the length of SL; (3) width of the space HJ between the two quadrifoils.

$$Ans. \ (1) \ \frac{r}{4}\sqrt{2}; \ (2) \ \frac{r}{4}\sqrt{2}; \ (3) \ \frac{r}{4}(2-\sqrt{2}).$$

6. If $OE = r$, find the area of the following figures: (*a*) the quadrifoil $ELFMG$ etc.; (*b*) the cusped figure ELF; (*c*) the quadrifoil $PQRS$ etc.; (*d*) the space between the boundaries of the two quadrifoils.

$$Ans. \ (a) \ \frac{r^2}{2}(\pi+2); \ (b) \ \frac{r^2}{8}(\pi-2); \ (c) \ \frac{r^2}{8}(3\pi+4); \ (d) \ \frac{r^2}{8}(\pi+4).$$

7. What per cent of the area of quadrifoil $ELFMG$ etc. is occupied by the inner quadrifoil? *Ans.* About 65 %.

8. Construct the quadrifoils $ELFMG$ etc. and $PQRS$ etc. in the square $ABCD$ in the position shown in Fig. 147.

9. If $AB = a$, find the area of the quadrifoil $PQRS$ etc.

$$Ans. \ \frac{a^2}{32}(3\pi+4).$$

184. 1. In a given square construct a quadrifoil of circles that intersect at the center of the square, so that each circle shall be tangent to two sides of the square.

Suggestion. — See Fig. 150. Inscribe a circle in each of the triangles DAB, ABC, BCD, and CDA.

2. Prove that these circles intersect on the diameters HF and GE.

Suggestion. — In order to prove that Q, the intersection of circles X and Y, lies on GE, prove that (1) $OX = OY$; (2) OE is the perpendicular bisector of XY; (3) $XQ = QY$, and hence (4) Q lies on OE.

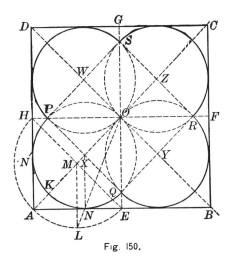

Fig. 150.

Fig. 150a. — Steel Ceiling Design.

3. Prove that the arc PKQ is a semicircle.

4. Prove that $PQRS$ is a square.

5. Morris in Practical Plane and Solid Geometry gives the following construction : On HE construct a semicircle. Draw ML, a radius perpendicular to AB. Draw LO meeting AB at N. Draw NX perpendicular to AB, meeting AO at X. Draw QX perpendicular to OA, meeting OH and OE at P and Q respectively. Prove that a semicircle constructed on PQ as diameter is tangent to the lines AH and AE.

Suggestion. — Compare the similar triangles OXQ and OME, also OXN and OML. Prove $XQ = XN = XO$.

6. If $AB = a$, find the length of OX. $\qquad$ *Ans.* $\dfrac{a}{2}(2 - \sqrt{2})$.

7. If $OX = b$, find the length of AB. $\qquad$ *Ans.* $b(2 + \sqrt{2})$.

8. If $AB = a$, find the area of the following figures : (1) the quadrifoil; (2) one of the ovals in the square $PQRS$; (3) the figure bounded by $\overset{\frown}{N'KN}$ and line-segments AN and AN'.

Ans. (1) $a^2(3 - 2\sqrt{2})(\pi + 2)$; (2) $\dfrac{a^2}{4}(3 - 2\sqrt{2})(\pi - 2)$;

$$(3)\ \dfrac{a}{8}(3 - 2\sqrt{2})(4 - \pi).$$

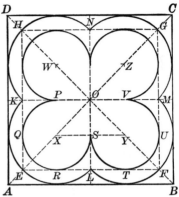

Fig. 151. — Carved Wood Design. Door, Rhyerson Laboratory, University of Chicago.

185. In Fig. 151 ABCD is a square. The quadrifoil KELFM etc. is constructed as in Fig. 150. It intersects the diagonals at the points E, F, G, and H, and the diameters at the points K, L, M, and N. The inner quadrifoil is placed as shown.

EXERCISES

1. Prove that the points E, L, and F are collinear and that $EFGH$ is a square.

2. Show how to construct the quadrifoil $PQRST$ etc.

3. Prove that points K, P, V, and M are collinear.

4. Prove that P is the middle point of KO.

5. Prove that points X, S, and Y are collinear.

6. If $AB = a$, find the length of OX and XS.

$$Ans. \text{ (1) } \frac{a}{2}(2 - \sqrt{2}); \text{ (2) } \frac{a}{2}(\sqrt{2} - 1).$$

7. If $AB = a$, find the area of the following figures: (a) the quadrifoil $PQRSTU$ etc.; (b) the quadrifoil $KELFM$ etc.; (c) the space between the boundaries of the two quadrifoils.

$$Ans. \text{ (a) } \frac{a^2}{4}(3 - 2\sqrt{2})(3\pi + 4); \text{ (b) } a^2(3 - 2\sqrt{2})(\pi + 2);$$

$$\text{(c) } \frac{a^2}{4}(3 - 2\sqrt{2})(\pi + 4).$$

8. What fraction of $ABCD$ is occupied by each quadrifoil? Find also the ratio of $EFGH$ to each quadrifoil.

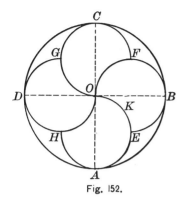

Fig. 152.

Fig. 152a. — Tiled Floor Design.
Scale ¾ in. = 1 ft.

186. 1. Give construction for Fig. 152.

2. Prove that the figures $OGCF$ and $OFBEK$ are congruent.

3. If $OB = r$, find the area of the following figures: (*a*) $OFBEK$; (*b*) $OFBAEKO$; (*c*) $OFBADHO$.

$$Ans. \ (a) \ \frac{r^2}{8}(\pi + 2)\ ; \ (b) \ \frac{\pi r^2}{4}; \ (c) \ \frac{\pi r^2}{2}.$$

4. Show how to divide a circle into four equal parts by means of arcs of circles.

Suggestion. — See Ex. 3*b*.

5. Show how to divide a circle into two equal parts by means of arcs of circles.

Suggestion. — See Ex. 3*c*.

6. Show how to divide a circle into any number of equal parts by means of arcs of circles. Use methods similar to those suggested in 4 and 5. See p. 178, note.

7. Make a drawing for one-fourth of the design shown in Fig. 152*a*.

8. How many degrees in the arc OKE?

Remark. — The figure drawn as suggested in Ex. 5 is the trademark of the Northern Pacific Railroad.

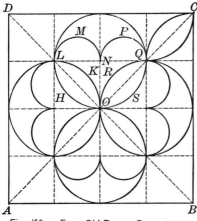

Fig. 153. — From Old Roman Pavement at
Brescia.   Gruner, Pl. 27.

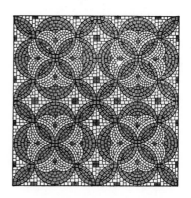

Fig. 153a. — Mosaic Floor Design.

EXERCISES

187. 1. Given the square $ABCD$, to construct the ornamental figure indicated in Fig. 153.

2. If $AB = a$, find the area of the curved figures (1) $ORQPNMLK$ and (2) $OSQPNMLH$.   *Ans.* (1) $\dfrac{a^2}{64}(8 - \pi)$; (2) $\dfrac{3\,a^2\pi}{64}$.

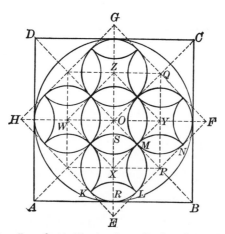

Fig. 154. — From Gothic Woodcarving.   Brochure Series, 1900, p. 113.

188. 1. Construct Fig. 154 within a given square.

2. If $AB = a$, find the length of XR and OP.

$$Ans. \ (1) \ \frac{a}{2}(\sqrt{2} - 1) \ ; \ (2) \ a(\sqrt{2} - 1).$$

3. If $AB = a$, find the area of the oval KL and of the four-pointed figure with the center at X.

$$Ans. \ (1) \ \frac{a^2}{8}(3 - 2\sqrt{2})(\pi - 2) \ ; \ (2) \ \frac{a^2}{4}(3 - 2\sqrt{2})(4 - \pi).$$

4. With the square $EFGH$ given as a basis construct Fig. 154.

5. If $EF = b$, find the length of OM and AB.

$$Ans. \ (1) \ \frac{b}{4} \ ; \ (2) \ \frac{b}{2}(\sqrt{2} + 1).$$

6. If $EF = b$, find the area of one of the ovals and of one of the four-pointed figures. $\qquad Ans. \ (1) \ \frac{b^2}{32}(\pi - 2) \ ; \ (2) \ \frac{b^2}{16}(4 - \pi).$

PART 4. ROUNDED MULTIFOILS

189. History and Occurrence. — The designs in this Part abound as large circular windows, as circular lights in pointed windows, and as leaded glass fillings for semicircular spaces over doors and windows.

Large circular windows are frequently met in England, but are so abundant in France as to form a striking characteristic of French Gothic architecture.[1] (See Figs. 162 and 183a.) As circular lights in pointed windows, the forms in this section seem to be somewhat characteristic of the earlier geometrical tracery.[2] For examples of the use of multifoils, see Figs. 139a, 164b, 166b, 189a, 193a, 200a, 210a, 239a, and 242a.

In **modern construction** designs may be found based on a ring of even seven, nine, or eleven small circles. Thus

---

[1] Fletcher, p. 380. [2] Bond, p. 472.

the Lutherkirche in Berlin has a window containing seven such circles, as has also the New Old South Church of Boston.   St. Georgenkirche of Berlin has windows containing nine and eleven circles.[1]

190.   Since medieval times these designs have been used for **inlaid tiles** (Fig. 164*a*).   In Jervaulx Abbey, Yorkshire, England, is a tiled pavement in which are designs based on rings of six and twelve tangent circles inside a given circle.   This dates from the thirteenth century.[2]

A very beautiful ceiling in the Camera del Doge, Ducal Palace at Venice, is made up of eight-foiled figures.[3]

Note. — In constructing many of these figures it is evident that since certain regular polygons, such as the 7-sided, 9-sided, 11-sided, 13-sided, etc., polygons cannot be constructed by the instruments ordinarily admitted in geometric construction, in practice this construction must be approximated by other means.   The problem is equivalent to the problem of dividing the perigon into a given number of equal parts.

FORMS BASED ON RINGS OF TANGENT CIRCLES

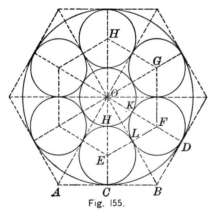

Fig. 155.   Fig. 155*a*. — From Chester Cathedral, England.

[1] Otzen, Vol. II, Pls. 1 and 4.   [2] Shaw (1), Pls. VII–XII.
[3] Brochure Series, Pl. XIX, 1898.

EXERCISES

191. **1.** Construct within a given circle a ring of six equal circles so that each circle shall be tangent to the given circle and to two of the small circles. See Fig. 155.

2. Prove that a seventh circle can be constructed tangent to each of these six circles and that its radius is equal to the radius of one of these six circles.

3. Prove that the radius of each of these circles is $\frac{1}{3}$ of the radius of the given circle.

4. If $OC = a$, find the area of the cusped figures, (*a*) HKL and (*b*) CLD.   *Ans.* (*a*) $\dfrac{a^2}{18}(2\sqrt{3} - \pi)$; (*b*) $\dfrac{a^2}{54}(5\pi - 6\sqrt{3})$.

5. Prove that the lines joining the centers of the circles E, F, G, etc. form a regular hexagon.

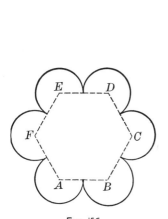

Fig. 156.

Fig. 156*a*. — Second Universalist Church, Boston, Massachusetts.

EXERCISES

192. **1.** Construct the rose window shown in Fig. 156 on the regular hexagon ABC etc.

2. If $AB = 5$, find the area of the window.

3. If $AB = a$, find the same area.   *Ans.* $\dfrac{a^2}{2}(3\sqrt{3} + 2\pi)$.

4. If $AB = a$, find the area of the circle that will circumscribe the window.   *Ans.* $\dfrac{9\,\pi a^2}{4}$.

5. Inscribe the rose window shown in Fig. 156 in a given circle.

6. If the radius of the circumscribed circle is a, find the area of the rose window,   *Ans.* $\dfrac{2\,a^2}{9}(3\sqrt{3} + 2\,\pi)$.

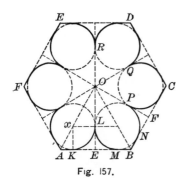

Fig. 157.

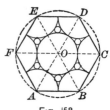

Fig. 158.

EXERCISES

193. 1. Construct the rosette shown in Fig. 157, given the radius of the regular hexagon.

Suggestion. — Construct the regular hexagon and the perpendiculars from the center on to the sides. In each of the quadrilaterals thus formed inscribe a circle.

2. If K is the point of contact of circle x and line AE, prove that $\dfrac{AK}{KE} = \dfrac{1}{\sqrt{3}}$.

3. If $AB = a$, find the radius of circle x.

4. If $AB = a$, find the areas of (*a*) the figure bounded by $\overset{\frown}{KL}$, $\overset{\frown}{LM}$, and the line-segment KM; (*b*) the entire rosette $AKLMBNP$ etc.; (*c*) the curved hexagon $LPQR$ etc.

EXERCISES

194. 1. Give construction for Fig. 158, if the hexagon *ABC* etc. is given.

2. If *AB* = *a*, find the radius of one of the small circles.

Suggestion. — Draw the medians of the triangle *AOB*. Use the theorem, " The medians of a triangle meet in a point whose distance is ⅔ the distance from the vertex to the opposite side."

3. If *AB* = *a*, find the area of the cusped hexagon, containing circle *O*.

195. In Fig. 159 AB is divided into three equal parts, AP, PQ, and QB, and semicircles are constructed on the line-segments, AB, AP, PQ, and QB, as diameters. Circles O and O' are tangent as shown.

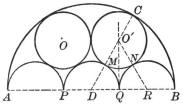

Fig. 159. — From Euclid Avenue Christian Church, Cleveland, Ohio.

EXERCISE

Find the radius of circles *O* and *O'*. *Ans.* $\dfrac{s}{6}$.

Suggestion. — Figure 159 is a portion of Fig. 155. Radius of circle *O'* may also be found by means of the equation $O'D = O'R$ or $\left(\dfrac{s}{2} - r\right)^2 = \left(\dfrac{s}{6} + r\right)^2$, where $s = AB$ and r is the radius of circle *O*.

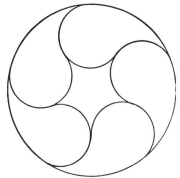

Fig. 160. — From Wetzel Building, East 44th Street, New York City. Brickbuilder, Nov. 1905.

Fig. 160*a*. — From Exeter Cathedral, England.

196. **1.** Show how a design like that in the circle in Fig. 160*a* may be derived from Fig. 155.

2. If the radius of the circle is *a*, find (1) the area of one of the lobed figures, and (2) the area of the cusped hexagon.

$$Ans. \ (1) \ \frac{a^2}{54}(11\pi - 6\sqrt{3}); \ (2) \ \frac{a^2}{9}(6\sqrt{3} - 2\pi).$$

3. Give the construction for Fig. 160.

4. Show that a circle may be inscribed in the cusped pentagon in the center of Fig. 160.

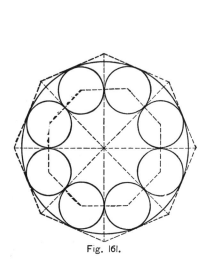

Fig. 161.

Fig. 161*a*. — From Monk's Church, St. Louis.

197. **1.** Construct within a given circle a ring of eight tangent circles, so placed that each circle is tangent to the given circle and to two small circles. See Fig. 161.

2. Show that the lines joining their centers form a regular octagon.

3. Show that a circle may be inscribed in the cusped octagon in the center of Fig. 161.

4. Show how to construct the eight-foiled figure shown in the circle in the window in Fig. 161*a*.

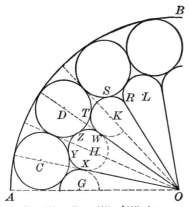

Fig. 162. — From Wheel Window.
Cathedral of Chartres, France.

Fig. 162a. — Leaded Glass Design.

198. Figure 162 shows one quarter of a wheel window which is composed of two rings of tangent circles. The outer ring is composed of circles tangent to the inclosing circle. Each circle of the inner ring is tangent to two of the outer ring.

EXERCISES

1. Show how to construct the outer ring of tangent circles, C, D, etc.

2. Show how to construct the inner ring of tangent circles.

Suggestion. — This involves the problem, " To construct a circle tangent to two given intersecting lines and to a given circle." See § 253.

3. If circle H is tangent to circles G and K at points X and W, respectively, find the number of degrees in the arc $XYZW$.

Ans. 202½.

Suggestion. — Solve by means of a pair of simultaneous equations. Let circle H be divided into two arcs a and b by the points X and W. The two equations are $a + b = 360$, $a - b = 45°$. Use the theorem about the measurement of an angle formed by two tangents.

4. Prove that the curved figures $OXYZW$ and $OWTSRO$ are congruent.

5. Draw a diagram for Fig. 162a. How are the leaf-shaped figures formed from the tangent circles?

Suggestion. — This involves the problem, " To find the center of a circle that shall pass through three given points."

199. Construct around a given circle a ring of any number of tangent circles, so that each circle shall be tangent to the given circle and to two of the small circles. See Fig. 163.

Note that this can be done only in case it is possible to divide the perigon into a number of equal parts corresponding to the number of small circles required. See § 190.

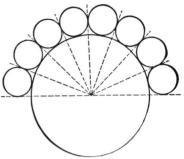

Fig. 163. — From Scroll Saw Pattern.

FORMS BASED ON SEMICIRCLES

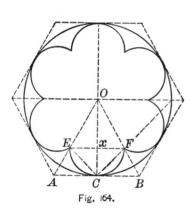

Fig. 164.

Fig. 164a. — Inlaid Tile Design.
1 in. = 1 ft.

200. 1. Construct within a given circle a rosette of six equal semicircles, so that each semicircle shall be tangent to the given circle. See Fig. 164.

Suggestion. — Circumscribe a regular hexagon about the given circle and draw radii to the points of tangency. In each triangle, as $\triangle AOB$, construct the right $\triangle ECF$ so that the vertex C shall be at the mid-point of AB and $OE = OF$.

2. Prove that any two successive semicircles intersect intervening radii at the same point.

3. Prove that the centers of these semicircles are equally distant from O.

4. If $OC = r$, find the length of Cx and the area of the rosette.

Ans. (1) $\dfrac{r}{2}(\sqrt{3} - 1)$;

(2) $3\,r^2\,(2 - \sqrt{3})(2\,\sqrt{3} + \pi)$.

5. If $EF = a$, find the area of the rosette and the length of CO.

Ans. (1) $\dfrac{3\,a^2}{4}(2\,\sqrt{3} + \pi)$;

(2) $\dfrac{a}{2}(\sqrt{3} + 1)$.

6. Inscribe in a given circle a rosette of semicircles of 8 parts; of 12 parts; of 16 parts; of 5 parts. See Fig. 164b.

Fig. 164a. — From Lincoln Cathedral, England.

FORMS BASED ON REVERSED CURVES

201. In Fig. 165, $\overparen{AMD}$, $\overparen{DLC}$, and $\overparen{CKB}$ are drawn from the middle points of $\overline{AD}$, $\overline{DC}$, and $\overline{CB}$, respectively, as centers, and are tangent to each other at the intersections of the chords GF and FE with the radii OD and OC.

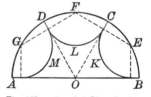

Fig. 165. — Leaded Glass Design.

EXERCISES

1. Show how to construct the figure.

2. If $OA = a$, find the areas of (a) the cusped figure in the center; (b) one of the oval figures. *Ans.* (a) $\dfrac{a^2}{4}(3\sqrt{3} - \pi)$; (b) $\dfrac{a^2}{4}(\pi - \sqrt{3})$.

3. Construct like figures dividing the circle into six, eight, or twelve equal parts.

4. Construct the figures mentioned in Ex. 3 with the same points as centers, but with radii equal to the chords of the half arcs.

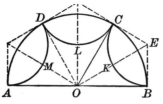

Fig. 166. — Leaded Glass Design.

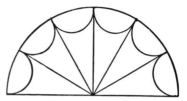

Fig. 166a. — Leaded Glass Design.

202. In Fig. 166 BCDA is a semicircle constructed on AB as diameter. $\overset{\frown}{BKC}$, $\overset{\frown}{CLD}$, and $\overset{\frown}{DMA}$ are tangent to each other at the points C and D.

EXERCISES

1. Show how to construct $\overset{\frown}{AMD}$, $\overset{\frown}{DLC}$, and $\overset{\frown}{CKB}$.

2. Prove that the figures $OBKCO$ and $OCLDO$ are congruent.

3. If $OB = r$, find the length of BE. $Ans.\ \dfrac{r}{3}\sqrt{3}.$

4. If $OB = r$, find the area of (a) $BKCDMA$ and (b) one oval.

$$Ans.\ (a)\ \frac{r^2}{3}(3\sqrt{3}-\pi);\ (b)\ \frac{r^2}{18}(5\pi-6\sqrt{3}).$$

Suggestion for (a). — Find the area of $OBECO$ and of the sector $ECKB$.

5. Construct such figures dividing the semi-circumference into six, eight, or twelve equal parts.

6. Construct a like figure in a circle dividing the circumference into sixteen equal parts. By the same method construct eight larger arcs inclosing the smaller ones in groups of two.

Remark. — The figure called for in Ex. 6 is the design of a Pompeian mosaic.[1]

Fig. 166b. — From St. Alban's Abbey Church, England.

203. In Fig. 167 the circumference of the circle is divided into six equal parts at the points A, B, C, D, E, and F. $\overset{\frown}{AYC}$, $\overset{\frown}{CUE}$, and $\overset{\frown}{EWA}$ are tangent to each other at the points A, C, and E, and $\overset{\frown}{BZD}$, $\overset{\frown}{DVF}$, and $\overset{\frown}{FXB}$ are tangent at the points B, D, and F.

[1] Zahn (2), Vol. I, Pl. 15.

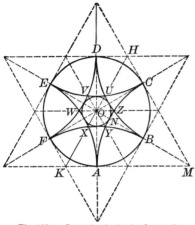

Fig 167. — From Diefenbach, Series B, Pl. XXXIII.

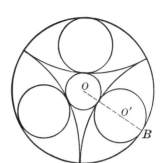

Fig. 167a. — From the Church of Our Lady of Lourdes, New York City. See Fig. 279a.

EXERCISES

1. Show how to construct $\overset{\frown}{A\,YC}$, $\overset{\frown}{C U E}$, and $\overset{\frown}{E W A}$.

2. Prove that the points K, X, U, and H are collinear.

3. Prove that

(a) $OX = OY = OZ$, etc.

(b) $\overset{\frown}{FX}$, $\overset{\frown}{XA}$, $\overset{\frown}{A\,Y}$, etc. are equal.

(c) $\overset{\frown}{XY}$, $\overset{\frown}{YZ}$, $\overset{\frown}{ZU}$, etc. are equal.

(d) It is possible to inscribe a circle in the curved hexagon XYZ etc.

4. If $AO = r$, find the radius of the circle tangent to XY, YZ, etc.

$Ans.\ r(2 - \sqrt{3}).$

5. If $AO = r$, find the following areas:

(a) The figure $ABCYA$.

$Ans.\ \dfrac{a^2}{6}(5\,\pi - 6\sqrt{3}).$

(b) The small circle with center O.

$Ans.\ \pi a^2(7 - 4\sqrt{3}).$

6. If $AO = r$, find the area of the three-pointed curved figure ACE.

$Ans.\ \dfrac{3\,r^2}{2}(2\sqrt{3} - \pi).$

7. Construct Fig. 167a, and find the radius of circle O' if $OB = r$.

$Ans.\ \dfrac{r}{2}(\sqrt{3} - 1).$

Suggestion. — See Fig. 167. $AM = r\sqrt{3}$; therefore, $NB = r(\sqrt{3} - 1)$.

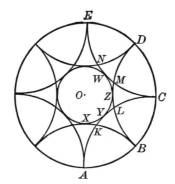

Fig. 168. — From Canopy over Portail des Libraires Cathedral, Rouen, France.   Porter, Vol. II,
p. 266.

EXERCISES

204. 1. Show how to construct Fig. 168 so that $\overarc{AYC}$, $\overarc{BZD}$, $\overarc{CWE}$, etc. shall be tangent at points A, B, C, etc.

2. Prove that the circle $XYZW$ can be drawn tangent to $\overarc{AYC}$, $\overarc{BZD}$, and $\overarc{CWE}$.

3. If $AO = r$, find the areas of (a) the oval $ABCY$; (b) the circle $XYZW$.   *Ans.* (a) $\dfrac{a^2}{2}(\pi - 2)$; (b) $\pi a^2(3 - 2\sqrt{2})$.

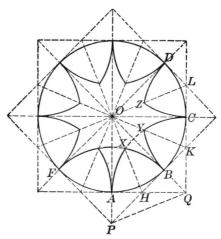

Fig. 169. — Detail from the Front of Strassburg Cathedral   Hartung, Vol. I, Pl. 21.

4. Construct a four-part figure like Fig. 167*a*.

Suggestion. — In Fig. 168 omit the arcs, *A YC*, *C WE*, etc., and inscribe circles in the four remaining ovals.

Note. — A window of this design occurs in Milan Cathedral.[1]

5. Find the area of the circle inscribed in one of these ovals.

$$Ans. \ \frac{\pi a^2(3 - 2\sqrt{2})}{2}.$$

EXERCISES

205. **1.** Give the construction for Fig. 169. Arcs *FXB*, *AXC*, *BZD*, etc. are tangent at points *A*, *B*, *C*, etc.

2. Prove that $\overparen{FXB}$ and $\overparen{AXC}$ intersect on the radius drawn to the midpoint of $\overparen{AB}$.

Suggestion. — Prove that *OH* extended is the perpendicular bisector of *PQ*. Since *PX = QX*, *X* lies on *OH*.

PART 5. OTHER FOILED FIGURES

206. The strong resemblance that many of the designs of this section bear to conventionalized flower forms suggests that as their origin. They have been widely used throughout the history of ornament.

SPECIAL ROSE WINDOW AND MOSAIC DESIGNS

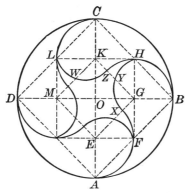

Fig. 170. — From D efenbach, Series B, Pl. XXX.

[1] Romussi, Pl. XXIV.

EXERCISES

207. 1. Show the construction for Fig. 170.

Suggestion. — Draw the two diameters AC and BD at right angles to each other. Connect the points as shown. F, H, etc. are the middle points of AB, BC, etc. E, G, etc. are the centers for $\overset{\frown}{AFX}$, $\overset{\frown}{BHZ}$, etc. $\overset{\frown}{XYH}$ is so drawn as to pass through the points X and H and have a radius equal to GB.

2. Prove that $\overset{\frown}{AFX}$ passes through the midpoint of the line-segment AB.

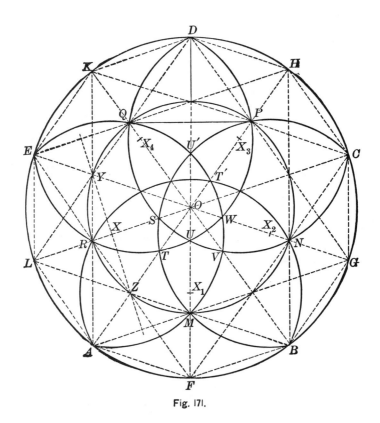

Fig. 171.

208. In Fig. 171 the 5-pointed $\frac{3}{10}$ star AMBNC, etc., is constructed in a given circle. Points X, X_1, X_2, etc. are centers for $\overline{E\,QT'MA}$, $\overline{ART'NB}$, etc.

EXERCISES

1. In the $\frac{3}{10}$ star shown in Fig. 171 prove that

(*a*) *EF*, *BK*, and *GD* are parallel.

(*b*) $EQ = QO = OR = RE = EL$.

(*c*) *DL*, *AK*, and *OE* are concurrent.

(*d*) *YZ* is the perpendicular bisector of *LO*.

(*e*) $LM = LO$.

(*f*) *X* is equally distant from points *E*, *Q*, *M*, and *A*.

2. Make a complete drawing for Fig. 171.

Fig. 171*a*. — From First Congregational Church, Chicago.

3. Prove that the points *A*, *Z*, *T*, *O*, *T'*, *P*, and *H* are collinear.

Suggestion. — This and many other properties of collinearity follow from the theorem : " If on a circle $\widehat{AB} = \widehat{BC}$ and if *A* and *C* are on opposite sides of a diameter through *B*, then circles with equal radii and with *A* and *C* as centers will intersect on the diameter through *B*, if they intersect at all." Prove this theorem. Show how it applies to prove that *P* lies on OX_3. How else does it apply in Ex. 3 ?

4. Prove that $OT = OV$, etc.

Suggestion. — Turn the figure through an angle of 72°. X_4 will fall on *X* and $\overline{ERPD}$ upon $\overline{AMQE}$. Show that point *T* will fall upon point *V*.

5. Prove that

(*a*) $OS = OU = OW$, etc.

(*b*) $TP = VQ$.

(*c*) $\widehat{RT} = \widehat{TM} = \widehat{MV}$, etc.

(*d*) $\widehat{TS} = \widehat{TU} = \widehat{UV}$, etc.

(*e*) $\widehat{SU} = \widehat{UW}$, etc.

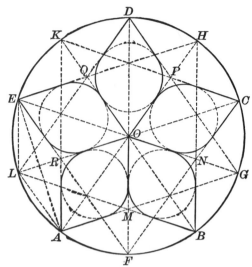

Fig. 172. — From Rose Window, Kent, England.   Brandon (I), Decorated, Sec. I, Pl. 15.

EXERCISES

209. 1. Make a complete drawing for Fig. 172.

2. If $OA = r$, find the length of (1) EL, (2) EA, and (3) the radius of one of the circles.

Ans. (1) $\dfrac{r}{2}(\sqrt{5} - 1)$; (2) $\dfrac{r}{2}\sqrt{10 - 2\sqrt{5}}$.

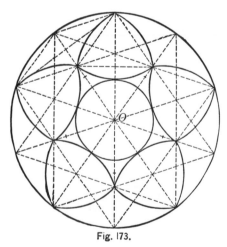

Fig. 173.

Fig. 173a.

EXERCISES

210. 1. Make a complete drawing for Fig. 173. Show that it depends directly upon Fig. 171. Why is the inner circle possible?

2. If the radius of the larger circle is r, find the radius of the inner circle.

Suggestion. — Use results obtained in § 209.

211. The dotted lines in Fig. 174 show a regular hexagon ABCDEF inscribed in a circle. G, H, K, L, M, and N are the middle points of the sides. These are joined as indicated.

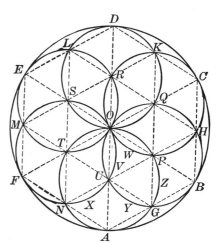

Fig. 174. — From a Pompeian Mosaic.
Zahn (I), Vol. II, Pl. 56.

EXERCISES

1. Prove that

(*a*) GK, HL, and OC are concurrent; (*b*) Q is the middle point of OC; (*c*) the hexagon is divided into a network of congruent equilateral triangles.

2. Show how to construct Fig. 174.

3. If $AO = r$, find the area of

(*a*) The segment bounded by $\overset{\frown}{PG}$ and the chord PG.

$$Ans. \ \frac{r^2}{48}(2\pi - 3\sqrt{3}).$$

(*b*) The oval OP.

$$Ans. \ \frac{r^2}{24}(2\pi - 3\sqrt{3}).$$

(*c*) The figure $UVOWP$.

$$Ans. \ \frac{r^2}{48}(9\sqrt{3} - 4\pi).$$

(d) The figure $UVOWPZGYU$.   *Ans.* $\dfrac{r^2}{8}\sqrt{3}$.

(e) The figure $ANXUYG$.   *Ans.* $\dfrac{r^2}{8}\sqrt{3}$.

(f) The entire curved figure $AGBHCK$ etc.

        Ans. $\dfrac{r^2}{4}(2\,\pi + 3\sqrt{3})$.

(g) The cusped figure AGB.   *Ans.* $\dfrac{r^2}{24}(2\,\pi - 3\sqrt{3})$.

4. If this design were to be executed in leaded glass, how many different shaped patterns would be necessary? Give proof.

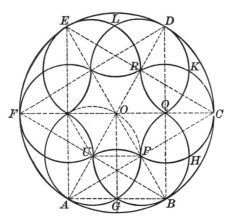

Fig 175. — From Gothic Wood-carving.
Brochure Series, July, 1900, p. 113.

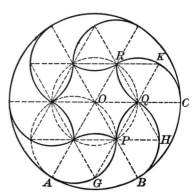

Fig. 175a. — Diefenbach,
Series B, Pl. XXX.

EXERCISES

212. 1. Show how to construct Figs. 175 and 175a.

Suggestion. — Use Fig. 174. G, H, K, etc. are the centers for UP, PQ, QR, etc.

2. If $OA = r$, find the area of (1) the curved triangle PHQ, and (2) the curved hexagon PQR, etc.

    Ans. (1) $\dfrac{r^2}{8}(\pi - \sqrt{3})$;   (2) $\dfrac{r^2}{4}(3\sqrt{3} - \pi)$.

3. In Fig. 175a prove that figures $AGPQHB$ and $BHQRKC$ are congruent.

4. If $AO = r$ in Fig. 175a, find the area of the figure $AGPQHB$.

$$Ans. \frac{r^2}{24}(5\pi - 3\sqrt{3}).$$

Suggestion. — This is $\frac{1}{6}$ of the difference between the whole circle and the curved hexagon PQR etc.

213. The dotted lines in Fig. 176 show ABC etc., a regular hexagon, inscribed in a circle, with alternate vertices joined as indicated.

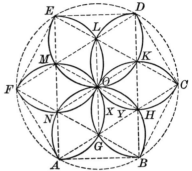

Fig. 176. — From Pompeian Mosaic. Zahn (2), Vol. III, Pl. 16. Also Vaulting Plan. " Handbuch der Architektur," II, 4, 2, p. 71, Pl., opposite p. 136.

EXERCISES

1. Prove that ⊿ ANG, NOG, GHB, etc., are congruent and equilateral.

2. Show how to construct the entire figure.

3. If $AO = r$, find the length of GO.

$$Ans. \frac{r}{3}\sqrt{3}.$$

4. If $AO = r$, find the area of

(*a*) The oval GO.

$$Ans. \frac{r^2}{18}(2\pi - 3\sqrt{3}).$$

(*b*) The figure $GXOYHB$.

$$Ans. \frac{r^2}{6}\sqrt{3}.$$

(*c*) The entire figure $AGBHC$ etc.

$$Ans. \frac{2\pi r^2}{3}.$$

(*d*) The cusped figure bounded by the arcs AG, GB, and AB.

$$Ans. \frac{\pi r^2}{18}.$$

5. What fraction of the hexagon $GHKL$ etc. is occupied by the ovals GO, HO, etc.?

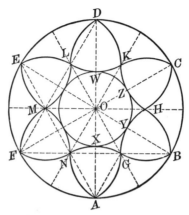

Fig. 177. — From an old Gothic Wood-carving. Brochure Series, July, 1900, p. 109.

EXERCISES

214. 1. Given the circle with the radius AO; construct Fig. 177.

Suggestion. — Use Fig. 176. Points A, B, C, etc. are the centers of $\overparen{NG}$, $\overparen{GH}$, $\overparen{HK}$, etc.

2. Construct the inner circle XYZ etc. tangent to the arcs NG, GH, HK, etc. Why is this possible?

3. If $AB = r$, find the area of (1) the curved triangle AGN; (2) the cusped hexagon GHK etc.; (3) the circle with radius OX.

$An_s.$ (1) $\dfrac{r^2}{6}(\pi - \sqrt{3})$; (2) $\dfrac{r^2}{3}(3\sqrt{3} - \pi)$; (3) $\dfrac{2}{3}\pi r^2(2 - \sqrt{3})$.

MANY–PARTED ROSETTES — TREATED AS TYPES

215. The designs in this group are treated as generalized types and are rich in possibilities. Four-part forms of the types shown in Figs. 180–182 are very common. One illustration is found in the window of the First Congregational Church, Oakland, Cal., which is based on Fig. 180. All-over designs made from this unit are common in Saracenic art. A very handsome one is found in the Hall of the Abencerrages in the Alhambra.[1] Win-

[1] Brochure Series, 1898, p. 133.

dows in Merton College Chapel, Oxford, England, contain this same unit.[1] In the sacristy of the Cathedral at San Miniato are frescoes containing four-part forms of the type shown in Fig. 181.[2] Similar eight-part forms are found in mosaics in St. Mark's, Venice.[3] These designs are also found in fancy needlework and in jewelry.

When designs of these types are executed in leaded glass, a pattern is first made for each piece, and the glass is then cut from the pattern. For many of the apparently complicated figures given in this paragraph the number of different patterns is comparatively few.

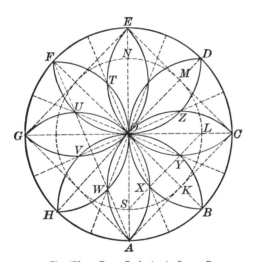

Fig. 178. — From Diefenbach, Series B, Pl. VIII.

Fig. 178a. — Stamped Steel Ceiling Design.

216. In Fig. 178 the circumference of the circle is divided into eight equal parts by the points A, B, C, D, etc. Point K on the diameter OB is the center for the arc CZOWA. The arcs intersect at the center and on the circumference as indicated.

[1] A. and A. W. Pugin, Vol. I, Pl. 4. [2] Waring (1), Pl. 13.
[3] Ongania, Large Folio, Pl. 10.

EXERCISES

1. Show how to determine point K on the radius OB so that $OK = KC = KA$. Why is this possible?

2. Prove that points X, O, and T are collinear.

3. Prove that the arc $CZOWA$ is a semicircle.

4. Prove that points X, Y, Z, etc. bisect the arcs BXO, AXO, BYO, CYO, etc.

5. Prove that (1) $OX = OY = OZ$, etc., and (2) arcs AX, XB, BY, etc. are equal.

6. Prove that chords joining every third point of division pass through two intersections.

Suggestion. — Prove that AF passes through points U and W.

7. Construct a diagram for the design shown in Fig. 178a and study its properties.

8. If this design is to be executed in leaded glass, how many different shaped patterns would be necessary?

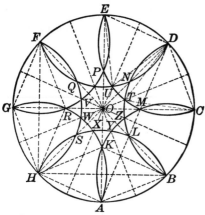

Fig. 179. — From Diefenbach, Series B, Pl. VIII.

Fig. 179a. — Ornamental Glass Design.

217. Figure 179 is an example of a class of figures that may be formed by dividing a circumference into any even number of parts by the points A, B, C, etc., and drawing arcs with these points as centers and AB as radius.

1. In Fig. 179 prove that

(*a*) points A, K, O, P, and E are collinear; (*b*) points X, O, and U are collinear; (*c*) line OX bisects the arc AH.

Suggestion. — See § 208, Ex. 3.

2. Draw like figures dividing the circle into twelve or sixteen equal parts and study the collinear points.

3. Draw a like figure dividing the circle into any odd number of equal parts, and study the collinear points.

4. Prove that distances KA, BL, CM, etc. are equal.

Suggestion. — Consider the figures $FGHR$ and $ABHK$.

5. Prove that if two pairs of equal circles intersect and the lines of centers are equal, the common chords are equal.

6. Prove that $OX = OY = OZ = OT$, etc.

Suggestion. — Rotate the figure through $\angle AOB$. Prove that OX and OY will coincide.

7. Prove that regular polygons may be drawn by joining the points K, L, M, N, etc., and W, X, Y, Z, etc., and that a circle may be constructed tangent to the arcs XY, YZ, ZT, etc.

8. Prove that the points A, L, M, and D are collinear.

9. Prove that the following arcs are equal:

(*a*) $SX = XK = KY = YL$.

Suggestion. — Describe a circle through the points S, K, L, M, etc. Draw chords SK, KL, etc., and SX, XK, KY, etc.

(*b*) $XY = YZ = ZT$, etc.

10. Draw like figures, dividing the circles into three, four, five, and six equal parts respectively, and study the properties of each. How is Fig. 179 constructed?

218. In Fig. 180 the circle is divided into twelve equal parts by the points A, G, B, H, C, etc. The points G, H, K, etc. are used as centers, and the chord of an arc of 90° as a radius, and the arcs drawn as indicated.

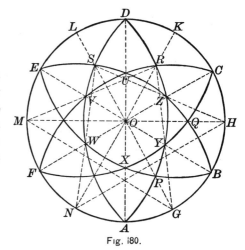

Fig. 180.

EXERCISES

1. Prove that points A, X, O, U, and D, also G, P, O, S, and L are collinear.

2. Prove that (*a*) $OP = OQ = OR$, etc., and (*b*) $OX = OY = OZ$, etc.

3. Prove that (1) $\widehat{AP} = \widehat{PB} = \widehat{BQ}$, etc.; (2) $\widehat{AY} = \widehat{YC} = \widehat{BZ}$, etc.; (3) $\widehat{XY} = \widehat{YZ} = \widehat{ZU}$, etc.

4. Prove that a circle can be inscribed in the curved figure $XYZU$ etc.

5. Construct such figures by dividing the circumference into four, five, eight, or twelve equal parts, and study their properties. In each case what points are used as centers for the arcs, if the radius is the chord of the arc of 90°?

6. If $AO = r$, find the area of

(*a*) The oval between the arcs $FVUC$ and $FXYC$.   *Ans.* $r^2(\pi - 2)$.

(*b*) The crescent-shaped figure between the arcs $FEDC$ and $FVUC$.

(*c*) The area of the circle inscribed in the figure XYZ, etc.

$$\textit{Ans. } \pi r^2(3 - 2\sqrt{2}).$$

219. 1. Construct Fig. 181.

Suggestion. — Divide the circle into eight equal parts by the points A, B, C, etc. On the radius AO take K_1 any point. With K_1 as center and K_1B as radius draw the arc HRB. Complete the figure.

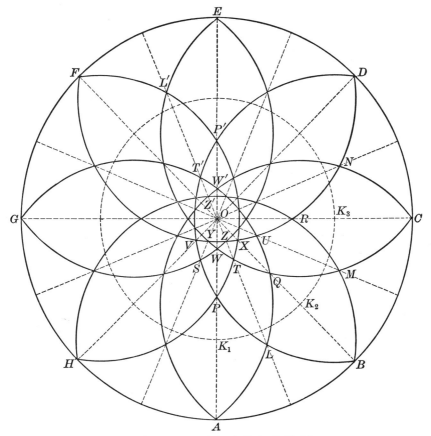

Fig. 181. — From Diefenbach. Series B, Pl. XX.

2. Prove that points A, P, W, O, W', P', and E, and also L, T, Z, O, Z', T', and L' are collinear.

3. What axes of symmetry has this figure?

4. Prove that

(*a*) $OL = OM = ON$, etc.

(*b*) $OP = OQ = OR$, etc.

(*c*) $OS = OT = OU$, etc.

(*d*) $OV = OW = OX$, etc.

(*e*) $OY = OZ$, etc.

Fig. 181a. — Iron Bracket.

Suggestion. — To prove each of these rotate the figure about the point O through an angle of 45°.

5. Prove the following:

(*a*) $\widehat{AL} = \widehat{LB} = \widehat{BM}$, etc.; (*b*) $\widehat{LQ} = \widehat{QM} = \widehat{MR}$, etc.; (*c*) $\widehat{PT} = \widehat{TQ} = \widehat{QU}$, etc.; (*d*) $\widehat{SW} = \widehat{WT} = \widehat{TX}$, etc.; (*e*) $\widehat{VY} = \widehat{WY} = \widehat{WZ}$, etc.

Suggestion. — From the symmetry with respect to line LL', prove $\widehat{AL} = \widehat{LB}$, etc. By rotation through an angle of 45° prove $\widehat{AL} = \widehat{BM}$, etc.

6. If this design is to be executed in leaded glass, how many different shaped patterns would be necessary? Give proof.

7. Construct such figures dividing the circumference into five, six, or any number of equal parts and study their properties.

220. In Fig. 182 the circle is divided into eight equal parts and radii drawn to the points of division. Lines CY and CW are constructed so that △ YCW is equilateral. The arcs YC and CW are drawn with W and Y as centers and WY as radius.

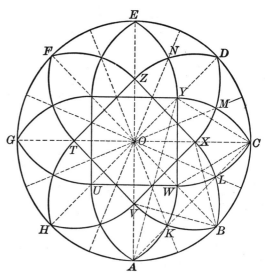

Fig. 182. — From Burn, p. 20.

EXERCISES

1. Construct the entire figure.

2. Prove that (*a*) $OK = OL = OM$, etc., and (*b*) $OV = OW = OX$.

3. Prove that the following arcs are equal: (*a*) $AW = WC = BX$, etc.; (*b*) $VK = KW = WL$, etc.

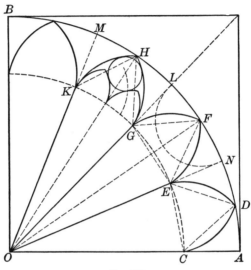

Fig. 183.

EXERCISES

212. 1. Given the circumference divided into any number of equal parts and radii drawn to the points of division, construct the equilateral △ CDE, EFG, etc. with the vertices D, F, H, etc. at the mid-points of the arcs AN, NL, etc., as shown in Fig. 183.

Suggestion. — The angles OFE, OFG, etc. must be 30°.

2. Prove that FE, DE, and ON are concurrent.

3. Prove that $OC = OE = OG = OK$.

4. Complete the construction of the figure.

Suggestion. — C, E, G, etc. are centers for the arcs ED, DC, FE, etc. For the details in figure KHG see § 275.

5. Prove that the curved figures $OCDE$ and $OEFG$ are congruent.

Remark. — Figure 183*a* shows a variation of the design in Fig. 183. The window in the north transept[1] of the same cathedral shows also a design related to Fig. 183.

Fig. 183*a.* — Window South Transept.
Notre Dame, Paris.

[1] Fletcher, Pl. 153.

CHAPTER V

GOTHIC TRACERY: POINTED FORMS

222. Classification. — Most of the figures in this and the two following sections are diagrams of tracery windows. **English tracery windows** have been divided into three classes : [1]

(1) Geometrical 1245–1315 A.D.
  (*a*) Earlier Geometrical.
  (*b*) Later Geometrical.
(2) Curvilinear 1315–1360 A.D.
(3) Rectilinear 1360–1500 A.D.

223. Characteristics. — The **earlier geometrical** designs are largely combinations of two units. The most common is shown in Fig. 229, which by repetition forms windows of four and eight lights. (See Fig. 231.) The other unit is of the type shown in Fig. 242 which by repetition forms windows of six lights. In these earlier designs the circles are filled with rounded trefoils, quadrifoils, and multifoils.

Later geometrical designs are characterized by trefoils and quadrifoils without the inclosing circle, pointed trefoils and curved triangles and squares. Windows of three,

[1] Sharpe (1), Chap. II.
205

five, and seven lights were common.   Designs of this character were developed in much greater variety in England and Germany than in France.

Most curvilinear designs, while constructed on a geometrical basis,[1] are highly complicated.   Very few are given here.   (See Figs. 196 and 197.)   A noted example is the beautiful east window in Carlisle Cathedral, which is made up of arcs of 263 different circles.[2]   These forms were highly developed in France where the style is known as the Flamboyant.

The **rectilinear** designs,[3] otherwise known as the perpendicular or Tudor, are particularly English.

Circular forms (Figs. 159, 202, 233, 234, 240, and 241) are from Romanesque or Renaissance buildings.

PART 1.   FORMS BASED ON THE EQUILATERAL TRIANGLE

224.   The **equilateral triangle** is said to form the **basis of Gothic tracery.**[4]   Of the designs that might be used in illustration, only the simpler are here given.   Many of these, however, may be considered as typical, inasmuch as they occur and recur throughout the entire Gothic period, forming the basis for, or entering as details into, more complicated designs.   The analysis of the more complicated forms may be found in books on architecture. Other designs showing the use of the equilateral triangle in Gothic tracery are given in Figs. 204–208, 213, and 214.

---

[1] For an analysis of many of these, see Brandon (1).

[2] Bond, p. 507; Billings (1), Pls. 18 and 19.

[3] Fletcher, p. 380.   [4] Brandon (1), p. 40.

THE EQUILATERAL ARCH AND THE EQUILATERAL CURVED TRIANGLE

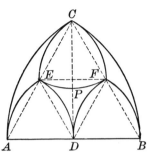

Fig. 184. — From the Door to the Chapter House, Wells Cathedral. Bond, p. 123.

EXERCISES

225. 1. Show how to construct an equilateral Gothic arch. See Fig. 184.

Suggestion. — Construct the equilateral triangle *ABC*. With *A* and *B* as centers and *AB* as radius construct the arcs *BC* and *AC*.

2. If *E*, *F*, and *D* are the middle points of the lines *AC*, *CB*, and *AB*, respectively, prove that equal equilateral triangles are formed.

3. Construct the equilateral arches *ADE* and *DBF* and the curved triangle *EFC*.

Suggestion. — Points *D*, *E*, *C*, *F*, *B*, and *A* are the centers and *AD* is the radius for the arcs drawn as indicated.

4. Prove that (1) arcs *EF*, *FD*, and *DE* are tangent in pairs; (2) arcs *AC* and *CE* are tangent; (3) line *CD* and arcs *ED* and *DF* are tangent.

5. If *AB* = 8, find the area of the arch *ABC*.

$$Ans. \ \frac{16}{3}(4\pi - 3\sqrt{3}).$$

Suggestion. — The area of $\triangle ABC$ is $16\sqrt{3}$. The area of the sector *ABC*, using *B* as the center of the circle, is $\frac{32\pi}{3}$. The area of the segment *AC* is $\frac{32\pi}{8} - 16\sqrt{3}$. Add the area of the segment to that of the sector *CAB*, using *A* as the center.

6. If $AB = 8$, find the area of (a) the arch AED;   (b) the curved figure CEF;   (c) the curved figure EFD;   (d) the curved figure CFB.

7. If $AB = s$, find the areas mentioned in Exs. 5 and 6.

Ans. (Ex. 5) $\dfrac{s^2}{12}(4\pi - 3\sqrt{3})$;   (b) $\dfrac{s^2}{8}(\pi - \sqrt{3})$;

(a) $\dfrac{s^2}{48}(4\pi - 3\sqrt{3})$;   (c) $\dfrac{s}{8}(2\sqrt{3} - \pi)$.

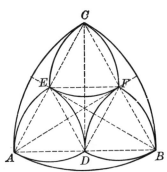

Fig. 185. — From the First Presbyterian Church, Salt Lake City, Utah.

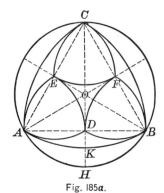

Fig. 185a.

EXERCISES

226.   1. Give the construction for Fig. 185.

2. Inscribe Fig. 185 in a given circle.   See Fig. 185a.

3. In Fig. 185a if $AO = r$, find the areas of the curved figures: (1) ABC; (2) ADE; (3) $AHBKA$; (4) $AKBDA$; (5) EDF.

Ans. (1) $\dfrac{3\,r^2}{2}(\pi - \sqrt{3})$; (3) $\dfrac{r^2}{6}(3\sqrt{3} - \pi)$;

(2) $\dfrac{3\,r^2}{8}(\pi - \sqrt{3})$; (4) $\dfrac{r^2}{8}(2\pi - 3\sqrt{3})$;

(5) $\dfrac{3\,r^2}{8}(2\sqrt{3} - \pi)$.

Fig. 185b. — From Carlisle Cathedral.

4. Construct the three figures AED, DBF, CEF in a given circle.

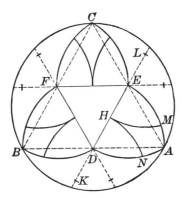

Fig. 186. — From Leighton, Pl. 39.

227. 1. Show the construction for Fig. 186.

Suggestion. — *E*, *D*, and *F* are the middle points of the sides of the equilateral triangle *ABC*. Arc *HN* has a radius equal to *BD* and has its center on line *ED* extended.

2. If the radius of the circle is *r*, find the area of (*a*) the figure bounded by $\widehat{AE}$ and $\widehat{AD}$ and the line-segment *DE*; (*b*) the figure *BDAECF*.

$$Ans. \ (a) \ \frac{r^2}{16}(4\,\pi - 3\,\sqrt{3}); \quad (b) \ \frac{3\,r^2}{8}(2\,\pi - \sqrt{3}).$$

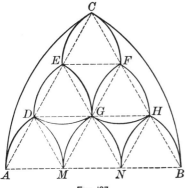

Fig. 187.

Fig. 187*a.* — From Exeter Cathedral.

228.   1. Give construction for the design shown in Fig. 187.

Suggestion. — Construct the equilateral $\triangle ABC$. Divide each side into three equal parts and join the points as indicated. The intersections are the centers and AM is the radius for the arcs drawn as indicated.

2. Pick out the pairs of tangent arcs and prove.

3. If this design is to be executed in leaded glass, how many different shaped patterns would be necessary? Give proof.

4. If $AB = 6$, find the areas of the curved figures: (a) ADM; (b) DMG; (c) $GFCE$; (d) $ADECA$; (e) EDG.

5. If $AB = s$, find the areas of the figures mentioned in Ex. 5.

Ans. (a) $\dfrac{s^2}{108}(4\pi - 3\sqrt{3})$; (b) $\dfrac{s^2}{18}(2\sqrt{3} - \pi)$; (c) $\dfrac{s^2}{18}\sqrt{3}$;

(d) $\dfrac{s^2}{18}(2\pi - 3\sqrt{3})$; (e) $\dfrac{s^2}{18}(\pi - \sqrt{3})$.

6. Construct the design shown in Fig. 187a by dividing each side of an equilateral triangle into five equal parts.

7. If each side of the triangle shown in Fig. 187a is a, show that the area of one of the curved quadrilaterals is $\dfrac{a^2}{50}\sqrt{3}$.

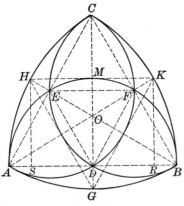

Fig. 188.

Fig. 188a. — Fourth Presbyterian Church, Chicago.

229. 1. Construct Fig. 188.

Suggestion. — Construct the equilateral $\triangle ABC$. A, B, and C are the centers for $\overset{\frown}{CB}$, $\overset{\frown}{CA}$, and $\overset{\frown}{AB}$. The semicircles are constructed on the sides of the linear triangle as diameters.

2. Prove that the semicircles intersect in pairs on the sides of the triangle.

3. If BE and AF extended intersect the arcs AC and BC at H and K, respectively, prove that HK is parallel to AB and tangent to $\overset{\frown}{AMB}$.

Suggestion. — Construct HS and KR perpendicular to AB. Prove $HS = KR = DB$. To prove $KR = DB$ notice that $DB = FB$; prove $FB = KR$ by use of $\triangle AKR$ and AFB.

4. If $AB = s$, find the areas of the curved figures: (a) DFE; (b) AEC; (c) AED.

$$Ans.\ (a)\ \frac{s^2}{8}(\pi - \sqrt{3})\,;\ (b)\ \frac{s^2}{24}(2\,\pi - 3\sqrt{3})\,;\ (c)\ \frac{s^2\pi}{24}.$$

5. If $AB = r$, find the lengths of CM and MK.

$$Ans.\ CM = MK = \frac{r}{2}(\sqrt{3} - 1).$$

6. Prove that $EFHK$ is a trapezoid and find its area if $AB = 10$.

$$Ans.\ \frac{25}{4}(5\sqrt{3} - 8).$$

7. Find this area if $AB = r$.

$$Ans.\ \frac{r^2}{16}(5\sqrt{3} - 8).$$

8. If this figure is to be executed in leaded glass, how many different patterns will be necessary? Give proof.

9. Has the figure a center of symmetry? An axis of symmetry? How many axes of symmetry?

10. If $AB = r$, find the area of $\triangle HKG$.

11. What part of $\triangle HKG$ is $\triangle EFD$?

12. If $AB = r$, find the area of the oval $ADFEA$.

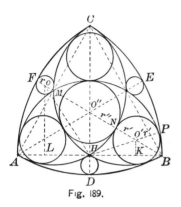

Fig. 189.

Fig 189.1. — First Congregational
Church, Chicago.

230. **Figure 189 is based upon Fig. 188.** Circles are inscribed in each of the seven curved triangles formed.

EXERCISES

1. If AE, BF, and CD are medians of the linear triangle ABC, and if they intersect at O'', prove that (1) O'' is the center of the circle inscribed in the inner curved $\triangle$, and (2) the medians pass through the points of tangency of circle O'' with the sides of the curved $\triangle$.

Suggestion. — See § 248, Ex. 1.

2. If $AB = s$, find the radius of circle O''.   *Ans.* $\dfrac{s}{6}(3 - \sqrt{3})$.

Suggestion. — $BM = \dfrac{s}{2}\sqrt{3}$, $NB = \dfrac{s}{2}(\sqrt{3} - 1)$, $O''N = O''B - NB$.
Notice that $O''B = \frac{2}{3} MB$.

3. Show how to construct circle O' tangent to the arcs as shown.

Suggestion. — Construct a line tangent to arc ANC at point N. Prove that if circle O' is tangent to this line at N, it is also tangent to the arc ANC at N. Then use § 246.

4. Show that if circle O' is tangent to each of the arcs HB and BM, its center must lie on the line BM.

5. Find the radius of circle O'.   *Ans.* $\dfrac{s(\sqrt{3} - 1)}{8 - 2\sqrt{3}}$.

Suggestion. — In $\triangle HO'K$, $\overline{O'H}^2 = \overline{O'K}^2 + \overline{HK}^2$, $O'H = \left(\dfrac{s}{2} - r'\right)$,

$$O'K = \dfrac{1}{2} O'B = \dfrac{BN - r}{2} = \dfrac{s\sqrt{3}}{4} - \dfrac{s}{4} - \dfrac{r}{2}. \qquad HK = \dfrac{s}{2} - KB = \dfrac{s}{2} - \dfrac{\sqrt{3}}{2} O'B$$

$$= \dfrac{s}{2} - \dfrac{\sqrt{3}}{2}\left(\dfrac{s\sqrt{3}}{2} - \dfrac{s}{2} - r\right) = \dfrac{s}{2} - \dfrac{3s}{4} + \dfrac{s\sqrt{3}}{4} + \dfrac{r\sqrt{3}}{2}. \qquad \text{Substitute in}$$

$\overline{O'H}^2 = \overline{O'K}^2 + \overline{HK}^2$ and solve for r.

6. Find the radius of circle O.

Suggestion. — Use $\triangle BLO$ and HLO and proceed as in Ex. 5.

TREFOILED ARCHES, POINTED TREFOILS, AND CUSPS

231. Occurrence. — The designs that follow are from all stages in the development of tracery. **Trefoiled arches** and **pointed trefoils** occur in large numbers and great variety. (See Figs. 190, 192, § 234, Exs. 4–9.) The former are among the earlier Gothic forms,[1] the latter are characteristic of later geometric tracery.[2] Figures 194–195 give methods for constructing the earlier **cusps.** These were made of separate pieces of stone inserted on the under side of the tracery bars. Later they were real continuations of the moldings.[3] Methods for constructing cusped triangles and arches may be obtained from §§ 232–235.

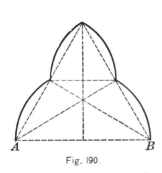

Fig. 190.

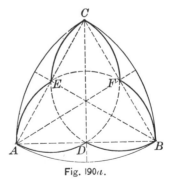

Fig. 190*a.*

[1] Parker (2), pp. 107–110, Illustrations.   [2] Sharpe (1), p. 73.

[3] Parker (2), p. 140.

EXERCISES

232. 1. Construct Fig. 190 and find its area if $AB = s$.

$$Ans. \ \frac{\pi s^2}{6}.$$

2. Construct Fig. 190a and find the area of $ADBFCE$ if $AB = s$.

$$Ans. \ \frac{s^2}{8} \ (2 \ \pi - \sqrt{3}).$$

3. Has Fig. 190 a center of symmetry? An axis of symmetry? More than one axis of symmetry?

4. Has Fig. 190a a center of symmetry? An axis of symmetry? More than one axis of symmetry?

5. Construct the trefoil $ADBFCE$, Fig. 190a, without the inclosing curved triangle.

Fig. 190b. — From First Presbyterian Church, Boston.

233. In Fig. 191 ABC is an equilateral curved triangle constructed as in Fig. 188. G, H, and K, the middle points of the arcs AB, AC, and BC, respectively, are the centers for the arcs AEB, CFA, and CEB, respectively.

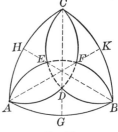

Fig. 191.

EXERCISES

1. Prove that $\overarc{AEB}$, $\overarc{CFA}$, and $\overarc{CEB}$ intersect on the medians of the linear triangle ABC.

2. Prove that the curved figures $AECHA$ and $AGBDA$ are congruent; also the figures $ADFEA$, $CEDFC$, and $BFEDB$.

3. Construct Fig. 191, omitting all dotted lines and the arcs AGB, AD, and DB.

Remark. — The resulting figure may be used in the heads of lights in large windows. The arcs AC and BC may also be omitted as in Fig. 190, forming a trefoiled arch.

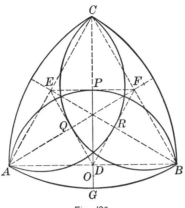

Fig. 192.

234. 1. Let E and F be the midpoints of CA and CB. Construct a figure like Fig. 190a, with the arc APB tangent to EF. See Fig. 192.

Suggestion. — Find point O on line CG equally distant from P and B.

2. Find the length of DO. *Ans.* $\dfrac{PD}{6}$.

Suggestion. — Let $AB = s$, and $DO = x$, $PD = \dfrac{s}{4}\sqrt{3}$, $DB = \dfrac{s}{2}$

Draw BO. Since PO is to equal BO, we get the equation

$$\left(\frac{s}{4}\sqrt{3} + x\right)^2 = \left(\frac{s}{2}\right)^2 + x^2.$$

3. If the centers for the arcs APB, BQC, CRA are arbitrary points on lines CD, BE, and AF, prove that the arcs intersect on the medians of the $\triangle ABC$, provided that the figure is symmetrical.

4. Construct several figures to illustrate Ex. 3.

5. If these figures are to be executed in leaded glass, how many different patterns are necessary in each case? Give proof.

6. Show how to construct trefoils similar to those obtained in Ex. 4 by means of the axes AF, BE, and CD without the curved $\triangle ABC$.

7. Show how Fig. 192 may be used to construct cusps in an equilateral arch.

8. Construct a trefoiled arch from Fig. 192.

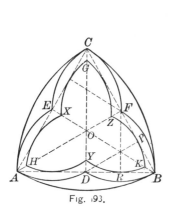

Fig. 193.

Fig. 193a. — From Church of Our Redeemer, St. Louis.

235. In Fig. 193 the trefoil ADBFC etc. is constructed from semicircles described on the sides of △ ABC. The points D, E, and F are the centers for the arcs which are tangent to the sides of △ ABC and which form the trefoil HYKZGX.

EXERCISES

1. If PD and RF are radii for the arcs ZK and YK, prove that $PD = FR$.

2. Prove that (1) PD, RF, and OB are concurrent; (2) ZK and YK intersect on OB; (3) GZ and KZ intersect on AF.

3. If $AB = s$, find the length of PD. $Ans. \dfrac{s}{4}\sqrt{3}.$

4. Join E and F and prove that the completed arc $HXZK$ would be tangent to EF.

5. Is it possible by the methods of elementary algebra and geometry to find the area of the inner trefoil?

Remark. — This figure often occurs without the inclosing curved triangle ABC.

236. In Fig. 194 ABC is an equilateral triangle. D, F, and E are the middle points of the sides AB, BC, and AC, respectively. K, L, and H, the middle points of OA, OB, and OC, respectively, are the centers for the arcs that form the cusps shown.

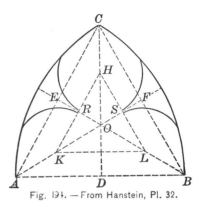

Fig. 194. — From Hanstein, Pl. 32.

EXERCISES

1. Prove that (1) *HKL* is an equilateral triangle, and (2) *BE* is the perpendicular bisector of *HK*.

2. Prove that the arcs that form the cusps are tangent at points *R* and *S*, the middle points of the lines *HK* and *HL*.

3. Prove that the circle inscribed in △ *ABC* will pass through the points *H*, *K*, and *L*.

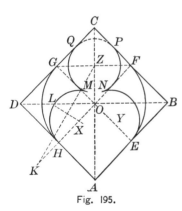

Fig. 195.

Fig. 195a. — Stairway, Rouen Cathedral.

237. In Fig. 195 ABCD is a square with diagonals AC and BD, and the diameters EG and HF. The circle MNPQ is inscribed in the square OFCG. The arcs GH and FE are quadrants with O as center, and OH as radius. The arc HLM is tangent to line AD at H, and to the circle MNPQ.

Construct the design shown in Fig. 195.

Suggestion. — For the construction of a circle tangent to a given circle and to a given line at a given point, see § 119, Ex. 7 or § 246.

CURVILINEAR FORMS

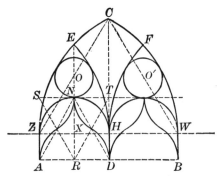

Fig. 196. — From St. Margaret's Church, Hertfordshire, Eng.   Brandon (I), Decorated, Sect. I, Pl. 13.

EXERCISES

238. 1. Given AB the span of given length. Construct the design for window shown in Fig. 196, making $\triangle ZCW$ equilateral.

Suggestion. — On the half span AD construct the equilateral $\triangle AND$ and its altitude NR. Make $NX = AR$. Through X draw ZW parallel to AB. AZ and WB are perpendicular to AB from A and B. The points Z and W are centers for the arcs CW and CZ. HZ is extended to a point G making $GH = ZW$, and G is center for the arc EH. Arc FH is similarly drawn. Arc ZNH is a semicircle on ZH as diameter. For circles O and O' see § 246. The remainder of the construction is indicated in the figure.

2. If $AB = a$, find the length of CH.

3. Find the radius of a circle which would be tangent to the arcs CE, CF, EH, and HF. See method used in § 281.

4. Find the radius of circle O.

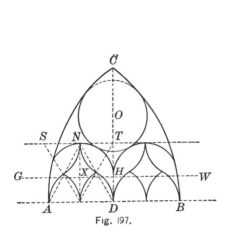

Fig. 197.

Fig. 197*a*. — From Trinity Church, Boston, Mass. Originally in St. Botolph's Church. Boston, Eng.

EXERCISES

239. 1. Make a diagram for Fig. 197*a*.

Suggestion. — See Fig. 197. The diagram is similar to that shown in Fig. 196. The radius for $\overset{\frown}{AC}$ and $\overset{\frown}{CB}$ is $\frac{5}{4}$ AB and the centers lie on AB extended. X is the center and NX the radius for semicircle through N and H.

2. Find the radius of circle O tangent to $\overset{\frown}{AC}$, $\overset{\frown}{CB}$ and to the semicircles as shown, if $AB = s$.

Suggestion. — Solve for x in the equation,

$$\sqrt{\left(\frac{5s}{4} - x\right)^2 - \left(\frac{3s}{4}\right)^2} = \sqrt{\left(\frac{s}{4} + x\right)^2 - \left(\frac{s}{4}\right)^2} + \frac{s}{4}\sqrt{3} - \frac{s}{4}.$$

PART 2. FORMS CONTAINING TANGENT CIRCLES: GEOMETRIC CONSTRUCTIONS

THE CIRCLE INSCRIBED IN THE GOTHIC ARCH.

240. Occurrence. — The diagrams given in Figs. 198 and 199 have been universally used since the Middle Ages. Almost every town can furnish one or more examples. A good illustration from St. Louis is given in Fig. 138*a* and one from Boston in Fig. 190*b*.

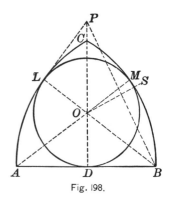

Fig. 198.

241. In Fig. 198 ACB is an equilateral arch with A and B as centers, and the circle with center O is tangent to the sides of the curved triangle at points D, M, and L.

EXERCISES

1. Prove that (1) $AD = BD$; (2) the sets of points C, O, D; B, O, L; A, O, M are collinear.

Suggestion. — Use the theorems (1) " If two equal circles intersect, any point in the common chord may be the center of a circle tangent to the given circles "; and (2) " If two circles are tangent, the line of centers passes through the point of tangency."

2. Construct the circle inscribed in the circular $\triangle ABC$.

Suggestion. — This involves the problem, to construct a circle tangent to a given line at a given point and to a given circle. Three methods are possible.

(*a*) An adaptation of that given in § 119, Ex. 7. Make $DP = AB$. Draw PB, and OS the perpendicular bisector of PB, meeting PD at O. Draw BO meeting the arc AC at L. Prove $OP = OB$; $PD = BL$. $\therefore OD = OL$, and the circle tangent to AC at L. Prove $OL = OM$.

(*b*) The method given in § 246.

(*c*) The following special method.[1] Make $PD = AB$. With P as center and DB as radius cut the arc AC at L. Draw LB cutting CD at O. Prove (1) $\angle PLB = $ rt. $\angle$, using $\triangle PLB$ and PDB; (2) $OL = OD$, using $\triangle PLO$ and ODB; (3) $OL = OM$.

3. If $AB = s$, find the length of DO. *Ans.* $\dfrac{3\,s}{8}$.

Suggestion. — Let $DO = x$. Since $LO = DO$, $BO = (s - x)$. Get the equation $(s - x)^2 - x^2 = \left(\dfrac{s}{2}\right)^2$.

---

[1] Hanstein, Pl. 19.

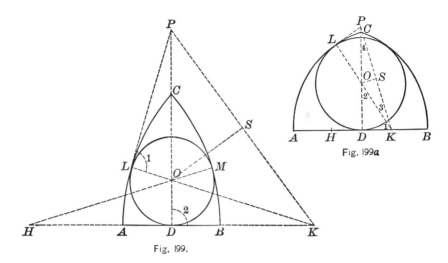

Fig. 199*a*

Fig. 199.

242. Figures 199 and 199*a* represent circles inscribed in Gothic arches which are not equilateral. Figure 199 shows the lancet arch and Fig. 199*a* the drop arch. In each case points H and K are the centers for the arcs BC and AC, respectively.

EXERCISES

1. Show how to inscribe a circle in a lancet arch ABC. See Fig. 199.

Suggestion. — Either method given in § 241, Ex. 2 may be used. For method (*c*) make $DP = AK$ and $PL = DK$. Draw LK meeting PD at O.

2. Show how to inscribe a circle in a drop arch ABC. See Fig. 199*a*.

The following exercises hold good for either figure.

3. If $AB = 8$ and $KA = 20$, find DO.   *Ans.* 3.6.

Suggestion. — Let $LO = x$. $\therefore KO = 20 - x$. Get the equation
$$(20 - x)^2 - x^2 = 256.$$

4. If $KA = ks$ and $AB = s$, find the length of DO.

$$Ans. \ \frac{s}{8}\left(\frac{4\,k - 1}{k}\right).$$

Suggestion. — Get the equation $(ks - x)^2 - x^2 = (ks - \tfrac{1}{2}s)^2$.

5. If $\dfrac{KA}{KD} = \dfrac{\sqrt{3}}{1}$, show that $DO = \frac{1}{3}KA$.

Suggestion. — If $\dfrac{PD}{KD} = \dfrac{\sqrt{3}}{1}$, $\angle PKD = 60°$, $\angle DPK = 30°$, $\angle DKO = 30°$.

6. If $DO = \dfrac{AB}{4}$, find AK.   *Ans.* $\dfrac{AB}{2}$.

Draw a figure to illustrate this case.

Suggestion. — To find AK, solve the equation $\dfrac{s}{8}\left(\dfrac{4k-1}{k}\right) = \dfrac{s}{4}$.

7. If $DO = \dfrac{AB}{n}$, find AK.   *Ans.* $\dfrac{ns}{4n-8}$.

8. Show that for all real arches n must be greater than 2.

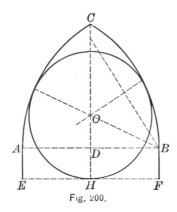

Fig. 200.

Fig. 200a. — From St. Anne's R. C.
Church, Chicago.

243. Figure 200 shows a stilted equilateral arch ECF with the inscribed circle. △ABC is equilateral. Points A and B are the centers for the arcs BC and AC, respectively. Lines AE and BF are equal straight lines perpendicular to AB.

EXERCISES

1. Show how to inscribe a circle in the arch EFC.

2. If $DH = \frac{1}{4}CD$, find the relation between AB and CH.

Ans. $\dfrac{8\sqrt{3}}{15}$.

3. If $AB = 12$ and $CH = 14\frac{1}{2}$, find the approximate length of DH.

Ans. 4.107.

4. If $AB = s$, $DH = h$, and $h < \dfrac{s}{2}$, find the radius of circle O.

$$Ans. \ \frac{3\,s^2 - 4\,h^2}{8(s - h)}.$$

Suggestion. — In $\triangle\, OBD$, $OB = s - r$, $BD = \dfrac{s}{2}$, $OD = r - h$. Solve the equation $(s - r)^2 = \dfrac{s^2}{4} + (r - h)^2$.

5. If $AB = s$, $DH = h$, and $h \gtreqless \dfrac{s}{2}$, find the radius of circle O.

$$Ans. \ \frac{s}{2}.$$

6. If $DH = \frac{1}{4}\,CD$, find the radius of circle O.

$$Ans. \ \frac{45\,s}{976}(8 + \sqrt{3}).$$

Suggestion. — Let $h = \dfrac{s}{8}\sqrt{3}$ in the formula found in Ex. 4.

7. Make a drawing for the window shown in Fig. 200*a*.

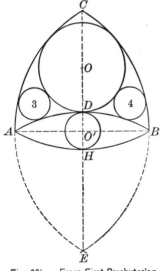

244. In Fig. 201 the line AB is given, and the arcs CAE and CBE are drawn with points B and A as centers and the arcs ADB and AHB with points E and C as centers, respectively.

Fig. 201. — From First Presbyterian Church, Fort Wayne, Indiana.

1. Construct circle O' tangent to arcs ADB and AHB.

2. Construct circle O tangent to arcs AC, BC, and ADB.

3. What loci must be used to find the centers of circles 3 and 4?

Suggestion. — The locus of the centers of circles tangent to any two given circles. This problem does not come within the province of elementary geometry.

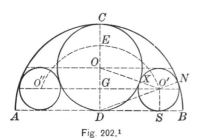

Fig. 202.[1]

Fig. 202a. — St. Andrews M. E. Church, New York City.

245. In Fig. 202 D is the middle point of AB. ACB is a semicircle. Circle O is tangent to the line AB and to the semicircle ABC. Circles O' and O'' are tangent, as indicated.

1. Show how to construct circle O.

2. If circle O' is tangent to circle O at X and to the semicircle ACB at N, prove that O, X, O' and also D, O', N are collinear.

3. If $AB = s$, find the radius of circle O' tangent to circle O, to the semicircle ACB, and to line AB.   *Ans.* $\frac{s}{8}$.

Suggestion. — Drop perpendiculars from O' to AB and CD. In $\triangle O'DS$, $\overline{DS}^2 = \overline{DO'}^2 - \overline{O'S}^2$. In $\triangle O'GO$, $\overline{O'G}^2 = \overline{OO'}^2 - \overline{OG}^2$. Since $DS = O'G$, $\overline{DO'}^2 - \overline{O'S}^2 = \overline{OO'}^2 - \overline{OG}^2$.

If radius of circle O' is r, $O'D = \dfrac{s}{2} - r$. $OO' = \dfrac{s}{4} + r$. $OG = \dfrac{s}{4} - r$.

The above equation becomes $\left(\dfrac{s}{2} - r\right)^2 - r^2 = \left(\dfrac{s}{4} + r\right)^2 - \left(\dfrac{s}{4} - r\right)^2$.

[1] Hanstein, Pl. 19.

4. Construct circle O'.

5. What fraction of the area of the semicircle ACB is occupied by the three included circles? *Ans.* $\frac{3}{4}$.

Remark. — Figure 202a shows a design related to Fig. 202. The construction of the two small circles on either side of the oval does not come within the province of elementary geometry.

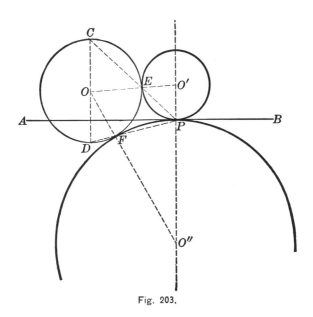

Fig. 203.

EXERCISES

246. **1.** PROBLEM. — To construct a circle tangent to a given line at a given point and to a given circle.

Suggestion. — See Fig. 203. Let O be the given circle, AB the given line, and P the given point. From O and P draw lines perpendicular to AB. Let the perpendicular through O cut the circle in points C and D. Draw CP cutting the circle at E. Draw OE meeting the perpendicular through P at O'. Prove $\triangle COE$ and PEO' similar. Hence, $O'E = O'P$, since $EO = OC$. Show that a second circle is possible. Give construction and proof for circle O''.

2. In the preceding construction show the application of the theorem, " If two circles have a common point in the line of centers, the circles are tangent."

3. What locus is used in this construction, and where?

4. Perform this construction in case point *P* lies within circle *O*.

5. For another solution for this problem, see § 119, Ex. 7.

CIRCLES INSCRIBED IN EQUILATERAL CURVED TRIANGLES

247. Occurrence. — The equilateral circular triangle occurs throughout Gothic architecture. It was used for small windows at an early date.[1] Fine examples are found in Westminster Abbey, where the inscribed circle contains an eight-foil, in Hereford Cathedral, where it contains a six-foil, and in Lichfield Cathedral (Fig. 208). That in Westminster Abbey dates from 1250. When used as details of tracery, circular triangles are an indication of later geometrical work. Its modern use is indicated in Figs. 188*a*, 189*a*, 204*a*, 205*a*, 206*a*, 258*a*.

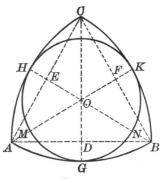

Fig. 204.

Fig. 204*a*. — From Second Universalist Church, Boston.

[1] Parker (2), pp. 139–140; Rickman, pp. 121, 190–192; Sharpe (1), p. 71.

EXERCISES

248. **1**. If ABC is an equilateral curved triangle with A, B, and C the centers for the arcs, show how to inscribe the circle in the triangle, as shown in Fig. 204.

Suggestion. — Let DC, EB, and AF be the medians of the linear triangle ABC. Extend them until they meet the arcs at G, H, and K, respectively. Let them intersect at O. O is the center for the required circle. Prove $OG = OH = OK$, and that the circle is tangent to the arcs AB, BC, and AC.

2. If CG extended is considered the locus of a certain class of circles tangent to each of the two given equal circles of which AC and CB are arcs, prove by intersecting loci that point O is the center of a circle tangent to the three arcs AB, AC, and CB.

3. If $AB = s$, find the length of GO and the area of the circle.

$$Ans. \ (1) \ \frac{s}{3}(3 - \sqrt{3}); \ (2) \ \frac{2\pi s^2}{3}(2 - \sqrt{3}).$$

4. If $GO = a$, find AB. $\qquad\qquad Ans. \ \frac{a}{2}(3 + \sqrt{3}).$

5. Prove that the curved triangles $AHMG$ and $BGNK$ are congruent.

6. If $AB = s$, find the areas of the curved triangles ABC and AGH.

$$Ans. \ (1) \ \frac{s^2}{2}(\pi - \sqrt{3}); \ (2) \ \frac{s^2}{18}(4\pi\sqrt{3} - 5\pi - 3\sqrt{3}).$$

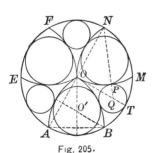

Fig. 205.

Fig. 205a. — From Boyton Church, England.

EXERCISES

249. 1. If the circumference of a circle is divided into six equal parts at points A, B, M, N, F, and E, show that it is possible to construct the equilateral curved triangles OAB, OMN, and OEF, with centers at these points as shown in Fig. 205.

2. If circle Q is constructed tangent to arcs OM, OB, and BM, prove that its center must lie on the radius perpendicular to AN.

3. Find the length of QT if $OT = r$.   *Ans.* $\dfrac{r}{4}$.

4. Give the complete construction of Fig. 205.

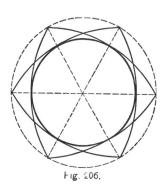

Fig. 206.

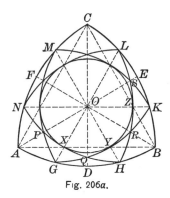

Fig. 206a.

EXERCISES

250. 1. Give complete construction for Fig. 206.

2. Show that Fig. 206 can be inscribed within an equilateral curved triangle ABC, if A, B, and C are the centers for the arcs BC, AC, and AB, respectively.

Suggestion. — In Fig. 206a ABC is an equilateral triangle with the medians AE, BF, and CD. The angles COF, FOA, AOD, etc., are bisected by the lines OM, ON, OG, etc. Prove that $OM = ON = OG$. Use superposition, folding along any median. Then prove that linear ⧍ GKM and HLN are equilateral and congruent.

Fig. 206b. — From St Bride's R. C. Church, Chicago.

3. Construct the curved ⚹ *GKM* and *HLN* and prove that a circle can be inscribed in the curved hexagon *XYZ* etc.

Suggestion. — Prove that $OX = OY = OZ$, etc., and that the circle and the arcs *NH*, *GK*, *HL*, etc., are tangent.

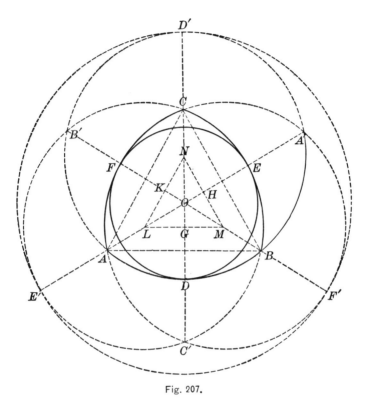

Fig. 207.

251. In Fig. 207 A*B*C is an equilateral triangle. L, M, and N are arbitrary points on the medians AE, BF and CD, respectively, so chosen that AL = BM = CN. L, M, and N are the centers and LC is the radius for the circles that intersect at A′, B′, and C′.

EXERCISES

1. Prove that these circles also intersect at points A, B, and C, and that the points A', B', and C' are on the medians extended.

Suggestion. — In order to prove that circles with centers at L and M intersect at point C prove that $LC = MC$,

2. Prove that a circle can be inscribed in the curved triangle ABC.

3. Show that Ex. 2 may be proved by intersecting loci.

4. Draw a diagram for one of the windows shown in Fig. 207a.

5. Draw a diagram like Fig. 207 with points L, M, and N on the lines EA, FB, and DC extended.

Fig. 207a. — St. Anne R. C. Church, Chicago.

6. Prove Exs. 1 and 2 for the figure called for in Ex. 5.

7. If $\triangle LMN$ is given equilateral and circles are constructed with L, M, and N as centers and with any equal radii, intersecting in points A, B, and C, and A', B', and C', prove that

(a) A circle can be inscribed in the curved $\triangle ABC$.

(b) A circle can be circumscribed about the three circles.

(c) The points E', A, O, A' are collinear.

(d) $AL = BM = CN$.

(e) Linear $\triangle ABC$ is equilateral.

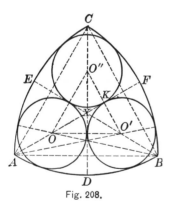

Fig. 208.

Fig. 208a. — From Lichfield Cathedral

252. In Fig. 208 ABC is an equilateral curved triangle with A, B, and C as centers for the sides. AF, BE, and CD are the medians of the linear $\triangle$ ABC, extended to the arcs as indicated. The circles O, O', and O'' are tangent to each other and to the sides of the triangle.

EXERCISES

1. If circle O is tangent to lines EX and XD, prove that its center must lie on line AX. If circle O is tangent to arcs AC and AB, prove that its center must lie on AX.

2. Construct circle O tangent to lines EX and DX and to the arcs AC and BC.

3. Make $O'B = OA$ and construct a circle with O' as center and a radius equal to that of circle O. Prove that this circle is tangent to lines XD and XF and to arcs BA and BC.

4. If $AB = s$, find the length of the radius of circle O'.

Ans. About $.24s$.

Suggestion. — In $\triangle AKO'$, $\overline{AK}^2 = \overline{AO'}^2 - \overline{KO'}^2$. If $OK' = x$, .
$XK = \dfrac{x}{3}\sqrt{3}$, and $AK = \dfrac{x}{3}\sqrt{3} + \dfrac{s}{3}\sqrt{3}$. Hence get the equation

$$(s-x)^2 - x^2 = \tfrac{1}{3}(s+x)^2.$$

5. What is the per cent of error if $O'K$ is called $\dfrac{s}{4}$?

EXERCISES

253. 1. Problem. — To construct a circle tangent to the sides of an angle and to a given circle.

Suggestion. — The solution may be accomplished by the three following steps.

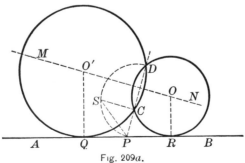

Fig. 209a.

(1) Construct a circle passing through two given points and tangent to a given line. See Fig. 209*a*.

Solution. — Let AB be the line and C and D the given points. Let CD intersect AB at P. Find the mean proportional between PC and PD. Lay off PQ and PR equal to the mean proportional. Erect perpendiculars to AB at Q and R. Draw the perpendicular bisector of DC, meeting the perpendiculars at O and O'. Prove that O and O' are the required circles.

Note that the problem is impossible if the two points are on opposite sides of the given line.

Study the case when one of the points is on the given line.

What locus is used in this construction?

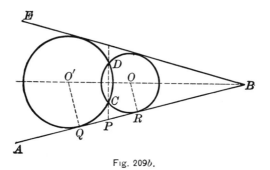

Fig. 209*b*.

(2) Construct a circle tangent to the sides of an angle and passing through a given point. See Fig. 209*b*.

Solution. — Let EBA be the given angle and C the given point. Draw $O'B$ the bisector of the angle. Draw CD so that $O'B$ is the perpendicular bisector of CD. Construct a circle passing through points C and D and tangent to AB. Prove that O and O' are the required circles.

Note that the problem is impossible if the given point is outside of the given angle.

When is only one circle possible?

What locus is used in this construction?

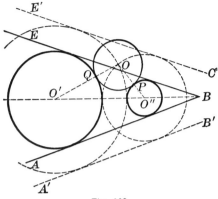

Fig. 209.

(3) *The main problem.* — To construct a circle tangent to the sides of angle *EBA* and tangent to circle *O*.

Solution. — Draw $A'B'$ and $E'C'$ parallel to the sides of $\angle$ *ABE*, and at a distance from them equal to the radius of circle *O*. Construct circle O' tangent to $A'B'$ and $E'C'$, and passing through point *O*. The circle with O' as center, and $O'Q$ as radius, is the required circle.

2. Study the following special cases:

(*a*) When the given circle cuts both sides of angle *EBA*.

Suggestion. — The parallels $A'B'$ and $E'C'$ may be inside the given angle. In this case four circles are possible.

(*b*) When it cuts one side of the angle and is·tangent to the other.

(*c*) When it cuts one side of the angle only.

(*d*) When it is wholly within the angle.

(*e*) When it is wholly without the angle.

(*f*) When it is tangent to one side only and within the given angle.

(*g*) When it is tangent to one side only and without the given angle.

(*h*) When it is tangent to both sides of the given angle.

TWO-LIGHT WINDOWS

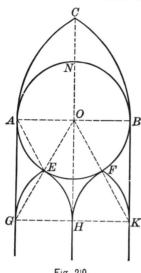

Fig. 210.

Fig. 210*a*. — First Presbyterian
Church, Chicago.

254. In Fig. 210 the arch ACB is equilateral, and the circle O is constructed on AB as diameter. The equilateral arches GEH and HFK are drawn with the half span AO as radius and with $\overparen{HE}$ and $\overparen{HF}$ tangent to circle O.

EXERCISES

1. Show how to find points *G*, *H*, and *K*, the centers for the arcs *EG*, *EH*, *FH*, and *FK*.

Suggestion. — This involves the problem, "To construct a circle of given radius, tangent to a given circle and to a given line." See § 265. Notice that the arcs are tangent to the lines *AG*, *CH*, and *BK*, as indicated.

2. How many circles can be constructed that will fulfill the conditions of the figure? Is there any difficulty in determining which circle is intended?

Suggestion. — See § 265.

3. Prove that the circle *O* and the arc *EH* are tangent at the apex of the arch *GEH*.

Suggestion. — Draw *GO* and *KO*. Prove that the line *GO* passes through the point of contact and that △ *GOK* is equilateral. Since ∠*EGH* = 60°, line *EG* falls on *OG*.

4. If $AB = 8$, find the area of the curved figures: (a) $\triangle HEF$; (b) $\triangle GEA$; (c) $\triangle ACBNA$.

Ans. (a) $8(2\sqrt{3} - \pi)$; (c) $\frac{8}{3}(5\pi - 6\sqrt{3})$.

5. If $AB = s$, find the area of the figures mentioned in Ex. 4.

Ans. (a) $\frac{s^2}{8}(2\sqrt{3} - \pi)$; (b) $\frac{s^2}{48}(9\sqrt{3} - 4\pi)$; (c) $\frac{s^2}{24}(5\pi - 6\sqrt{3})$.

6. If $AB = s$, find the length of AG. *Ans.* $\frac{s}{2}\sqrt{3}$.

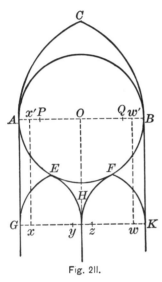

Fig. 2ll.

Fig. 2lla. — From Notre Dame, Paris.

255. Figure 211 shows a design like that of Fig. 210, with AQ and PB for radii of $\overparen{AC}$ and $\overparen{CB}$, respectively, each $\frac{5}{6}$ of AB, and the radii for $\overparen{EG}$, $\overparen{EH}$, $\overparen{HF}$, and $\overparen{FK}$, each $\frac{5}{6}$ of AO.

EXERCISES

1. Show how to construct the figure.

Suggestion. — Let Ox' and Ow' be $\frac{5}{6}$ of OA and OB, respectively. Draw $x'x$ and $w'w$ perpendicular to AB. The centers for arcs EH and FH must lie on these lines. Why? Determine points x and w the centers for the arcs EH and FH. Draw line wx and complete the construction.

2. Is $\triangle Oxw$ equilateral?

3. If $AB = s$, find the length of AG. *Ans.* $\frac{s}{3}\sqrt{6}$.

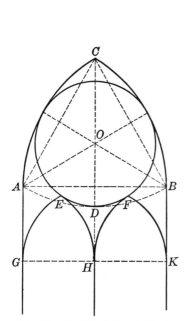

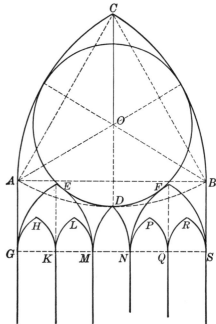

Fig. 212. — Northfleet Church, Kent,
England. Brandon, (I), p. 4I.

Fig 2I3. — From Chapter House, York Cathedral.
Rickman, p. I82.

256 In Fig. 212 the equilateral curved triangle ABC is constructed with A, B, and C as centers. The circle O is inscribed in this triangle. Arches GEH and HFK are equilateral, erected on the half span GH, so that the arcs HE and HF are tangent to circle O and to line CD extended.

EXERCISES

1. Construct the figure, making AB three inches. Show all work in detail.

Suggestion. — See § 248 and § 265.

2. Is the point of contact the summit of the arch GEH?

Suggestion. — Draw GO and OK. Is $\triangle GOK$ equilateral?

3. If $AB = a$, find the radius of circle O and the length of BK.

$$Ans. \ (1) \ \frac{a}{3}(3 - \sqrt{3}); \ (2) \ \frac{a}{3}\sqrt{21 - 9\sqrt{3}} - \frac{a}{6}\sqrt{3}.$$

257. In Fig. 213 the circle O is constructed as in Fig. 212. All arches are equilateral except MDN. Arcs EM and FN are tangent to circle O and constructed as in Fig. 210. Arcs MD and DN are constructed to pass through point D and be tangent to arcs EM and FN at points M and N respectively.

EXERCISE

Construct the figure, making AB three inches. Show all work in detail.

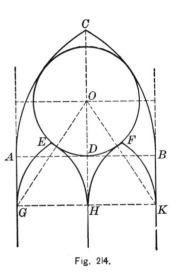

258. In Fig. 214 arch ACB is equilateral, with A and B as centers. Circle O is inscribed in the curved triangle ABC. Arches GEH and HFK are equilateral and tangent to circle O, as in Fig. 210.

Fig. 214.

EXERCISES

1. Construct a figure like Fig. 214, in which AB is five inches. Show all work in detail.

2. If $AB = s$, find the radius of circle O and the length of BK.

$$Ans. \ (1) \ \frac{3\,s}{8}; \ (2) \ \frac{s}{8}(\sqrt{33} - 3).$$

Suggestion. — For the radius of circle O see § 241, Ex. 3.

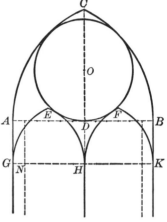

259. The design in Fig. 215 is like that in Fig. 214 with the radii of the arcs AC and CB each $\frac{5}{6}$ of the span AB, and the radii of the arcs EG, EH, HF, and FK each $\frac{5}{6}$ of AD.

Fig. 215. — From Lady Chapel, St. Patrick's Cathedral, New York City. Am. Arch. and Bldg. News, March 24, 1906.

EXERCISES

1. Construct a figure like Fig. 215 with $AB = 4$. Show all work in detail.

2. If $AB = s$, find the length of BK.   *Ans.* $.2935\,s$.

Suggestion. — For the radius of circle O, see § 242, Ex. 4.

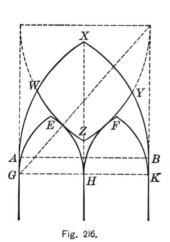

Fig. 216.

Fig. 216a. — From Cologne Cathedral.

260. In Fig. 216 the arch AXB and the curved square WXYZ are constructed as shown in § 115. The arches GEH and HFK are equilateral. The arcs EH and HF are tangent to the arcs WEZ and ZFY, respectively.

EXERCISES

1. Show how points G and K are determined, and prove GK parallel to AB.

2. Make a drawing for Fig. 216, showing all work in detail. Let AB be five inches.

3. If $AB = s$, find the length of BK.   *Ans.* $\dfrac{s}{2}(\sqrt{5} - 2)$.

4. Prove that X, Y, Z, and W are the vertices of a square.

5. Prove that Y is the midpoint of arc XB.

6. If $AB = a$, find the area of the square whose vertices are X, Y, Z, and W.

Suggestion. See § 115.

261. In Fig. 217 the circle is inscribed in the curved square constructed as in Fig. 216. In the equilateral arches GEH and HFK the arcs EH and FH are tangent to circle O.

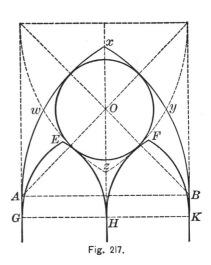

Fig. 217.

1. Construct a figure similar to Fig. 217, so that AB shall be four inches. Show all work in detail.

2. Prove that it is possible to inscribe circle O in the circular square $xyzw$.

3. If $AB = s$, find the radius of circle O and the length of BK.

$$Ans. \text{ (1) } \frac{s}{2}(2 - \sqrt{2}); \text{ (2) } \frac{s}{2}(\sqrt{10 - 6\sqrt{2}} - 1).$$

262. In Fig. 218 ACB is an equilateral arch. Point O is an arbitrary point on the altitude CD. The equilateral arches GEH and HFK are tangent to circle O.

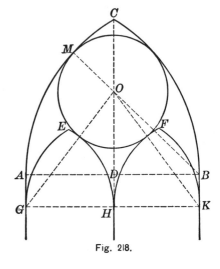

Fig. 2l8.

EXERCISES

1. Construct circle O tangent to the arcs AC and CB.

Suggestion. — Draw BO and extend to the arc AC. Use the theorem, " If two circles have a common point on the line of centers, the circles are tangent."

2. Construct the equilateral arches GEH and HFK on the half span AD and DB, so that the arcs EH and HF will be tangent to circle O.

3. If $AB = 12$ and $DO = 5$, find the length of OM and DH.

$$Ans. \ (1) \ 4.189 = (12 - \sqrt{61}) \ ; \ (2) \ 3.2.$$

4. If $AB = s$ and $DO = b$, find the length of OM and HO.

$$Ans. \ (1) \ s - \tfrac{1}{2}\sqrt{s^2 + 4\,b^2} \ ; \ (2) \ \tfrac{1}{2}\sqrt{9\,s^2 + 4\,b^2 - 6\,s\sqrt{s^2 + 4\,b^2}}.$$

5. If $DH = a$ and $AB = s$, find the length of DO.

263. Figure 219 shows a design like Fig. 218. $AQ = PB = \tfrac{5}{4} AB$. P and Q are the centers for the arcs CB and AC, respectively. O is an arbitrary point on the altitude CD. The arches GEH and HFK are tangent to circle O and drawn with radius $\tfrac{5}{4}$ of AD.

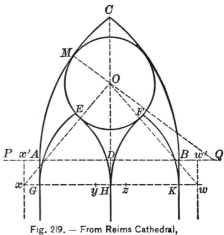

Fig. 2I9. — From Reims Cathedral, France, Sturgis (I), p. 267.

EXERCISES

1. Construct circle O tangent to the arcs CB and CA.

2. Construct arches GEH and HFK on the half span with a radius $\tfrac{5}{4}$ of AD, so that the arcs EH and HF are tangent to circle O and to the line CD extended.

Suggestion. — Find the points x, z, y, and w, the centers for the arcs EH, EG, FK, and FH, respectively.

3. If $AB = 8$ and $DO = 1\tfrac{3}{4}$, find OM and DH.

$$Ans. \ (1) \ 3^3 \ ; \ (2) \ \tfrac{1}{4} \ (5\sqrt{33} - 7).$$

4. If $AB = s$, $DO = b$, and $AQ = \frac{5}{4} AB$, find OM and OH.

Ans. (1) $\dfrac{5s - \sqrt{9 s^2 + 16 b^2}}{4}$; (2) $\frac{1}{4}\sqrt{59 s^2 + 16 b^2 - 15 s \sqrt{9 s^2 + 16 b^2}}$.

5. If $AB = s$, $AQ = ks$, and $DO = b$, find OM and OH.

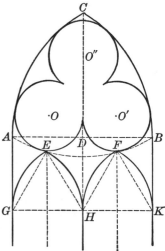

264. In Fig. 220 ABC is an equilateral curved triangle with A, B, and C as centers. The trefoil is inscribed in this triangle as in Fig. 208. Points E and F are obtained by dropping perpendiculars from the middle points of AD and DB. Arches GEH and HFK are equilateral arches on the half span GH = AD, passing through the points E and F.

Fig. 220. — From a German Church, Hartung, Vol. 2, Pl. 97.

EXERCISE

Show how to locate points G, H, and K.

Suggestion. — See § 266, Ex. 1.

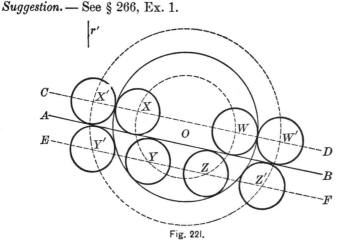

Fig. 221.

EXERCISES

265. PROBLEM. — Given the circle O with radius r and the line AB. To construct a circle of given radius r' tangent to circle O and to AB.

Suggestion. — Solve by intersecting loci. Use the locus of centers of circles of given radius and tangent to a given line, and the locus of centers of circles of given radius and tangent to a given circle. How many circles are possible?

1. What must be the distance d of line AB from center O in order that circle $X'Y'Z'W'$ may

(*a*) be tangent to line CD?   *Ans.* $d = 2r' + r$.

Suggestion for (*a*). — The radius of circle $X'Y'Z'W' = r' + r$, and the distance from O to CD is $d - r'$. In order that $X'Y'Z'W'$ may be tangent to CD, $r' + r$ must equal $d - r'$. Therefore, solve the equation $r' + r = d - r'$ for d.

(*b*) cut the line CD?   *Ans.* $2r' + r > d$.

(*c*) be tangent to line EF?

       Ans. $d = r$. AB is tangent to the given circle.

(*d*) cut line EF?   *Ans.* $d < r$. AB cuts the given circle.

2. If the radius of the required circle is less than the radius of the given circle, what must be the distance of line AB from the center O, in order that circle $XYZW$ may

(*a*) be tangent to line CD?

       Ans. $d = r$.   Line AB is tangent to the given circle.

Suggestion. — As in Ex. 1 the radius of circle $XYZW$ will be $r - r'$ and the distance of line CD from point O will be $d - r'$.

(*b*) cut line CD?   *Ans.* $d < r$.   The line AB will cut the circle.

(*c*) be tangent to EF?   *Ans.* $d = r - 2r'$.

(*d*) cut line EF?   *Ans.* $d < r - 2r'$.

3. Answer the questions in Ex. 2, if $r' > r$.

Ans. (*a*) $d = 2r' - r$; (*b*) $d < 2r' - r$; (*c*) $d = -r$; that is, AB is tangent to circle O but on the opposite side of the center from line EF; (*d*) $d < -r$; that is, AB is outside the circle.

4. If $2\,r' < r$, show that there will be

No solution to the given problem, if $d > 2\,r' + r$.
One solution, if $d = 2\,r' + r$.
Two solutions, if $2\,r' + r > d > r$.
Four solutions, if $d = r$.
Six solutions, if $r > d > r - 2\,r'$;
Seven solutions, if $d = r - 2\,r'$.
Eight solutions, if $d < r - 2\,r'$.

5. If $2\,r > r' > r$, show that there will be

No solutions to the given problem, if $d > 2\,r' + r$.
One solution, if $d = 2\,r' + r$.
Two solutions, if $2\,r' + r > d > 2\,r' - r$.
Three solutions, if $d = 2\,r' - r$.
Four solutions, if $d < 2\,r' - r$.

6. Discuss the cases (1) $2\,r' = r$, (2) $r' < r < 2\,r'$, (3) $r' = 2\,r$ (4) $r' > 2\,r$, (5) $r = r'$.

THREE–LIGHT WINDOWS

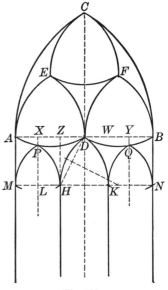

266. In Fig. 222 the equilateral arch **ACB** and the curved triangles are constructed as in § 226. The arches **MPH** and **KQN** are equilateral, constructed with $\frac{1}{3}$ **AB** as radius, so that their apexes will lie on the arcs **APD** and **DQB**, respectively. The arcs **DH** and **DK** are so constructed that their centers lie on the line **MN** and that they will pass through the points **H** and **D**, and **K** and **D**, respectively.

Fig. 222.

1. Show that the arc *PM* may be constructed by means of either of the following problems:

(1) To construct a circle passing through a given point, of given radius, and tangent to a given line.

(2) To construct a circle passing through a given point, of given radius, and with its center on a given line.

2. What loci would be used in each of the above constructions?

3. Show that the arc *DH* may be constructed by means of either of the following problems:

(1) To construct a circle passing through a given point and tangent to a given line at a given point.

(2) To construct a circle passing through two given points with its center on a given line.

4. What loci would be used in each of the above constructions?

267. In Fig. 223 ACB is an equilateral arch constructed from A and B as centers. AG = GH = HB. AGE, HBF, and EFC are equilateral curved triangles constructed from their vertices as centers. The arch RDS is also equilateral.

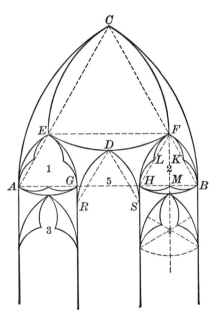

Fig. 223. — From St. Nazaire Cathedral, Carcassonne, France. Sturgis (I), p. 278.

EXERCISES

1. Show how to construct the three curved triangles AGE, HBF, and EFC from the linear triangle ABC.

2. If D is the center of the arc EDF, show how to locate points R and S so that $DR = RS = DS$. Construct the arcs DR and DS.

3. Construct a complete design like Fig. 223. Let AB be four inches.

Suggestion. — The trefoils in 1 and 2 are described from the middle points of the straight-line triangles as centers; those in 3 and 4 from the middle points of the sides of the circular triangles as centers. See §§ 232 and 233.

4. If $AB = s$, find the area of the curved figures: (a) the $\triangle CEF$; (b) the $\triangle CFB$; (c) the $\triangle KFB$; (d) the trefoil $HMBKF$ etc.

Ans. (a) $\dfrac{2\,s^2}{9}(\pi - \sqrt{3})$; ($b$) $\dfrac{s^2}{27}(2\,\pi - 3\sqrt{3})$; ($c$) $\dfrac{s^2}{216}(2\,\pi - 3\sqrt{3})$; ($d$) $\dfrac{s^2}{72}(2\,\pi - \sqrt{3})$.

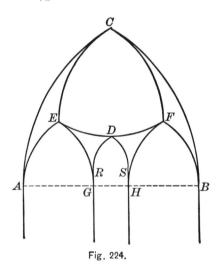

Fig. 224.

Fig. 224a. — From Exeter Cathedral.

EXERCISES

268. 1. Construct a design like that given in Fig. 223 with the span of arches 1 and 2 twice the span of arch 5.

2. If $AB = s$, find the area of the curved figures: (a) $\triangle CEF$; (b) $\triangle CFB$; (c) window head AEG; (d) window head RDS.

Ans. (a) $\dfrac{9\,s^2}{50}(\pi - \sqrt{3})$; (b) $\dfrac{s^2}{25}(2\,\pi - 3\sqrt{3})$; (c) $\dfrac{s^2}{75}(4\,\pi - 3\sqrt{3})$;

(d) $\dfrac{s^2}{300}(4\,\pi - 3\sqrt{3})$.

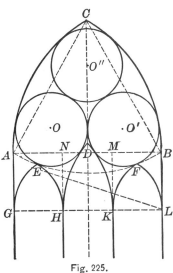

269. In Fig. 225 ABC is an equilateral curved triangle with A, B, and C as centers. Circles O, O', and O'' are constructed as in Fig. 208.

Fig. 225.

EXERCISES

1. If the equilateral arch GEH drawn with $\frac{1}{3}AB$ as radius is so constructed that arc EH is tangent to circle O, show how point G is determined.

Suggestion. — See § 265.

2. Determine points G and L so that arcs EH and FK, drawn with a radius equal to $\frac{1}{3}AB$ and with G and L as centers, will be tangent to circles O and O', respectively. Prove GL parallel to AB.

3. Let $AN = NM = MB$. Draw NH and MK perpendicular to AB from points N and M, respectively, meeting GL at H and K. Complete the construction of arches GEH and KFL.

4. If arc HD is tangent to line NH at point H and to circle O, how is the center for arc HD to be determined?

5. Construct arcs *DK* and *DH* by the same method and prove that they will have the same radii.

Fig. 226. — From Milan Cathedral.

270. In Fig. 226 the center of the large central circle is probably arbitrary. The two small side arches are constructed with one third of the main span as radius and with one side of each tangent to the circle. The small central arch is equilateral and made to touch the circle.

EXERCISES

1. What circle constructions are used for the small arches?

2. Make a diagram for a window like Fig. 226 with the span 4 inches and the radius of the sides of the main arch ⅝ of the span.

PART 3. FORMS CONTAINING TANGENT CIRCLES: ALGEBRAIC ANALYSIS

271. Most of the diagrams given in this Part are treated as types. By means of the formulæ given it is easy to obtain the data for any special cases under these types. The cut that accompanies the diagram is usually one of these possible special cases. Windows like those given and many others abound. The relation between the span and the radius of the arcs forming the main arch may be obtained roughly from even small photographs. If the length of the span and the altitude are known for any particular case,[1] the radius may be found as in § 275, Exs. 20 and 21. It is therefore possible to study with more or less accuracy any particular window desired.

ONE-LIGHT WINDOWS

272. Figures 228, 236, and 247 furnish illustrations of the following type of one-light window. The formula for their solution is given in Ex. 7 below.

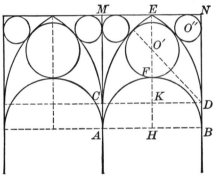

Fig. 227. — From Norwegian Legation, Washington, D. C. Arch. Annual, 1907.

In Fig. 227 CED is an equilateral arch constructed with C and D as centers. The rectangle CDNM is constructed with DN parallel to the altitude EK. ABDC is a rectangle with DB $= \frac{1}{4}$ AB. AFB is a semicircle on AB as diameter. Circles O′ and O″ are tangent as shown.

EXERCISES

1. Find the ratio of *DN* to *MN*.   *Ans.* $\dfrac{\sqrt{3}}{2}$.

2. If $AB = s$, find the radius of circle O'.   *Ans.* $\dfrac{11\,s}{40}$.

[1] E. Sharpe (2), gives the dimensions for a large number of windows from English churches and cathedrals.

Suggestion. — Use $\triangle DO'K$. $O'D = s - r$, $KD = \dfrac{s}{2}$, $O'K = \dfrac{s}{4} + r$.

Solve for r in the equation $(s - r)^2 - \dfrac{s^2}{4} = \left(\dfrac{s}{4} + r\right)^2$.

3. If $DB = hAB = hs$, find the radius of circle O'.

$$Ans. \ \frac{s}{2} \frac{(1 + 2\,h - 2\,h^2)}{3 - 2\,h}.$$

Suggestion. — In $\triangle DO'K$, $O'K = \left(\dfrac{s}{2} - hs + r\right)$.

4. Find the radius of circle O' if DB is zero. $Ans. \ \dfrac{s}{6}$.

Suggestion. — Let h be zero in the formula found in Ex. 3. See Fig. 228.

5. If $AMNB$ is a square,

(a) Find the ratio of $\dfrac{DB}{AB}$. $Ans. \ \dfrac{2 - \sqrt{3}}{2}$.

Suggestion. — If $NB = s$ and $ND = \dfrac{s}{2}\sqrt{3}$, $DB = \dfrac{s}{2}(2 - \sqrt{3})$.

(b) Find the radius of circle O'.

$$Ans. \ \frac{s}{8}(7 - 3\sqrt{3}) \text{ or about } \frac{9\,s}{40}.$$

6. Construct circle O''.

Suggestion. — See § 253 and § 332.

7. If the radius of $\overset{\frown}{CE}$ and $\overset{\frown}{DE}$ is ks and $DB = hs$, find the radius of circle O' when $AB = s$.

$$Ans. \ \frac{s}{2}\left(\frac{1 + 2\,h^2 - 2\,h - 2\,k}{2\,h - 2\,k - 1}\right).$$

8. Show how the formula obtained in Ex. 3 may be derived from that obtained in Ex. 7.

Suggestion. — Let $k = 1$. Why?

9. Show how the result in Ex. 2 may be derived from that in Ex. 7.

10. Draw a figure to illustrate the case mentioned in Ex. 5.

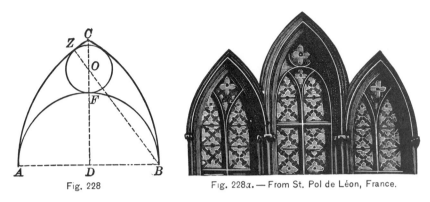

Fig. 228   Fig. 228a. — From St. Pol de Léon, France.

273. In Fig. 228 the diameter of the semicircle AFB is the span of the equilateral arch ACB. CD is the perpendicular bisector of AB. The circle O is tangent to the semicircle and to the arcs as indicated.

EXERCISES

1. Prove that the points B, O, and Z are collinear.

2. If $AB = s$, find the radius of circle O. $Ans. \dfrac{s}{6}.$

Suggestion. — $\triangle\, ODB$ gives the equation $(s - r)^2 = \left(\dfrac{s}{2}\right)^2 + \left(\dfrac{s}{2} + r\right)^2$ where r is the radius desired.

3. Construct circle O.

4. What per cent of the area above the line AB is occupied by the circle and semicircle?   *Ans.* About 78%.

5. Construct a figure like Fig. 228 in which the radius of $\overset{\frown}{AC}$ and $\overset{\frown}{CB}$ is $\frac{5}{6}$ of the span AB. Show that the radius of circle O in this figure is $\dfrac{AB}{8}$.

6. Find the radius of circle O, if the radius of $\overset{\frown}{AC}$ and $\overset{\frown}{CB}$ is ks.

$$Ans. \; \frac{s}{2} \cdot \frac{2\,k - 1}{2\,k + 1}.$$

7. Show that the formulæ obtained in Exs. 2 and 6 may be derived from that obtained in § 272, Ex. 7.

[1] Hanstein, Pl. 19.

TYPES OF TWO-LIGHT WINDOWS

274. Type I. — Figures 229–234, 237, 238, 256, and 258 furnish illustrations of this type. The formulæ for their solution are given in Exs. 9, 23, and 25 in § 275. The form of this type shown in Figs. 229 and 230 is the common design of early geometrical windows. Its use is universal. Figure 193a shows an illustration from St. Louis. By repetition of this design, windows of four and eight lights are formed as in Fig. 231. The ratio of the radius to the span varies greatly. It is said to be $\frac{3}{5}$ in the choir at Lincoln Cathedral, $\frac{5}{6}$ in Salisbury Cathedral, and $\frac{6}{5}$ in the nave at York Cathedral.[1]

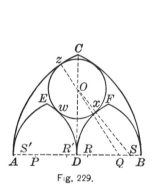

Fig. 229.

Fig. 229a. — From the Chapter House, Salisbury Cathedral.

275. In Fig. 229 AQ = BP = $\frac{5}{6}$ AB. P and Q are the centers for $\overarc{BC}$ and $\overarc{AC}$, respectively. The arches AED and DBF are constructed on the halves of the span AB, with radius $\frac{5}{6}$ AD and points S′, R′, R, and S as centers. Circle O is tangent to the arcs as shown.

[1] *Enc. Brit.*, Arches and Masonry in article on Building.

EXERCISES

1. Prove that (1) CD extended is the common chord of the two intersecting circles of which AC and BC are arcs; (2) CD is perpendicular to AB; (3) the linear $\triangle ACB$ is isosceles.

2. Prove that points z, O, Q, and also O, x, S are collinear, when z and x are the points of tangency of circle O and $\overset{\frown}{CA}$ and $\overset{\frown}{FD}$.

3. If circle O is tangent to the two equal intersecting circles AC and CB, or to the two equal tangent circles ED and FD, prove that its center lies on the line CD.

4. Find the radius of circle O, if $AB = s$.   *Ans.* $\dfrac{7\,s}{30}$

Suggestion. — Solve for r in the equation

$$\left(\frac{5\,s}{6} - r\right)^2 - \frac{s^2}{9} = \left(\frac{5\,s}{12} + r\right)^2 - \left(\frac{5\,s}{12}\right)^2.$$

5. Find the center of circle O.

Suggestion. — Use Q as center and $(AQ - r)$ as radius, and strike an arc cutting line CD at O.

6. If $AB = s$, what is the length of the radius used in Ex. 5 ?

Ans. $\dfrac{3\,s}{5}$.

7. If S is used as a center what radius must be used to obtain point O?   *Ans.* $SD + r$, or $\dfrac{13\,s}{20}$.

8. If $AQ = \frac{3}{5}\,AB$ and $DS = \frac{2}{5}\,DB$, find the radius of circle O, if $AB = s$.   *Ans.* $\dfrac{7\,s}{36}$.

9. If $AB = s$ and $AQ = ks$, find the radius of circle O.

Ans. $\dfrac{s}{12} \cdot \dfrac{4\,k - 1}{k}$.

Suggestion. — Let r be the radius of the circle. Then $OQ = ks - r$,

$QD = ks - \dfrac{s}{2}$, $\ OS = \dfrac{ks}{2} + r$, $\ DS = \dfrac{ks}{2}$.   Get the equation

$$(ks - r)^2 - \left(ks - \frac{s}{2}\right)^2 = \left(\frac{ks}{2} + r\right)^2 - \left(\frac{ks}{2}\right)^2.$$

10. If $AQ = \tfrac{3}{5} AB = \tfrac{3}{5} s$, find the radius of circle O by substituting $\tfrac{3}{5}$ for k in the formula obtained in Ex. 9.

$Ans.\ \dfrac{7\,s}{36}.$

11. If $AQ = \tfrac{1}{2} AB$, find the radius of circle O by substituting $\tfrac{1}{2}$ for k in the formula obtained in Ex. 9. Draw illustrative figure. See Fig 234.

$Ans.\ \dfrac{s}{6}.$

12. If $AQ = AB$, find the radius of circle O by substituting 1 for k in the formula obtained in Ex. 9. Draw illustrative figure. See Fig. 230.

$Ans.\ \dfrac{s}{4}.$

13. If $AQ = \tfrac{5}{6} AB$ and $AB = s$, find the radius of circle O by substituting in the formula obtained in Ex. 9.

$Ans.\ \dfrac{7\,s}{30}.$

14. If $AQ = ks$, and $Ox = r$, find value of AB in terms of k and r.

$Ans.\ \dfrac{12\,rk}{4\,k - 1}.$

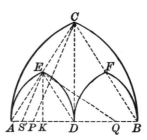

Fig. 229*b*.

15. If $AQ = \tfrac{5}{6} AB$ and $OX = 4$, find the lengths of AB, CD, EK, and DO. For EK see Fig. 229*b*.

$Ans.$ (1) $17\tfrac{1}{7}$; (2) $\tfrac{20}{7}\sqrt{21}$; (3) $\tfrac{10}{7}\sqrt{21}$; (4) $\tfrac{16}{7}\sqrt{14}$.

16. If $AQ = ks$, $S'D = k \cdot AD$, and $AB = s$, find the value of CD and EK. See Fig. 229*b*. $Ans.$ (1) $\dfrac{s}{2}\sqrt{4\,k - 1}$; (2) $\dfrac{s}{4}\sqrt{4\,k - 1}$.

17. Prove that $S'E$ and PC are parallel.

Suggestion. — From the computations in Ex. 16, $\triangle\ S'EK$ and PCD have their sides proportional and are therefore similar.

18. Prove that △ AED and ACB are similar.

Suggestion. — Since $S'E$ and CP are parallel, the isosceles △ $S'ED$, and PCB may be proved similar and $\angle CBP$ may be proved equal to $\angle EDS'$. For another discussion of this problem see § 328.

19. Prove that the points A, E, and C are collinear, and that E is the center of AC.

20. Find the value of AQ in terms of CD and AB.

$$Ans. \ AQ = \frac{4\,\overline{CD}^2 + \overline{AB}^2}{4\,AB}.$$

Suggestion. — Draw EQ. Prove △ ACD and AEQ similar. Hence $\frac{AQ}{AC} = \frac{AE}{AD}$.

21. If CD is 8 feet and AB 9 feet 10 inches, find the relation between AB and AQ.   *Ans. AQ is about $\frac{9}{10}$ of AB.*

22. Construct a figure in which $AQ = BP = \frac{5}{6}\,AB$, and the arches AED and DFB are equilateral. Show that the radius of circle O in this figure is $\dfrac{7\,s}{32}$.

23. If $AQ = k \cdot AB$ and $DS = h \cdot DB$, find the radius of circle O.

$$Ans. \ \frac{s}{4} \cdot \frac{4\,k - 1}{2\,k + h}.$$

24. Show that the formula obtained in Ex. 23 becomes that obtained in Ex. 9 if $h = k$.

Remark. — The center of circle O in Figs. 230, 232, and 234 may be found by the formula obtained in Ex. 23, as will be shown later.

25. Show that if the arch ACB is stilted by an amount equal to n, and if $AQ = k \cdot AB$ and $DS' = h \cdot DB$, the radius of circle O may be found from the equation, in which $AB = s$,

$$\sqrt{\left(ks - x\right)^2 - \left(ks - \frac{s}{2}\right)^2} = \sqrt{\left(h\frac{s}{2} + x\right)^2 - \left(h\frac{s}{2}\right)^2} - n.$$

Suggestion. — For an illustration of stilted arches see Fig. 233 in which AH represents the n of Ex. 25. Is there any limit to the value of n in a figure of the general type suggested in Ex. 25?

276. Figure 230 represents a special case under the preceding, and may be solved by formulæ obtained in § 275, Exs. 9 or 23. It is of common occurrence, and is more readily solved by methods suggested below.

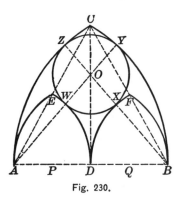

Fig. 230.

Fig. 230*a*. — From the Union Park Church, Chicago.

In Fig. 230 D is the middle point of AB, A is the center for the arcs CB and ED, B for the arcs AC and DF, and D for the arcs AE and BF. Circle O is tangent to the arcs as shown.

EXERCISES

1. Prove that $\overset{\frown}{ED}$ and $\overset{\frown}{DF}$, $\overset{\frown}{FB}$ and $\overset{\frown}{CB}$, $\overset{\frown}{AE}$ and $\overset{\frown}{AC}$ are tangent in pairs, and that CD extended is the common chord of the two intersecting circles of which AC and CB are arcs.

2. Prove that points A, E, C, also B, F, C are collinear.

3. If circle O is tangent to arcs AC and CB or to ED and FD, prove that its center must lie on the line CD.

Suggestion. — Use the theorem, " The perpendicular bisector of the line of centers of two equal circles is a part of the locus of the centers of circles tangent to the two given circles." What two cases of this theorem occur in this figure?

4. If circle O is tangent to the two concentric arcs ED and CB, prove that its center must lie on a circle drawn with A as a center and AQ as a radius, if Q is the middle point of the line DB.

5. If circle O is tangent to the two concentric arcs FD and AC, where must its center lie?

6. If circle O is tangent to the arcs CB, AC, ED, and DF at the points Y, Z, W, and X, respectively, prove that the points Z, O, X, B, and also Y, O, W, A are collinear.

Suggestion. — Use the theorem, " If two circles are tangent, the line of centers passes through the point of tangency."

7. Construct circle O.

Suggestion. — The construction may be performed by any two of three loci. It is necessary to make $OX = OW = OY = OZ$, and to prove the circle tangent to the arcs AC, BC, ED, and FD. The simplest proof is obtained by constructing the arcs drawn with A and B as centers and with AQ and BP as radii, respectively.

8. Show that the radius of circle O may be obtained from the formula in § 275, Ex. 9.

Suggestion. — Let $k = 1$ in the formula $\dfrac{s}{12} \cdot \dfrac{4k - 1}{k}$.

9. Show that the radius of circle O may be obtained from the formula in § 275, Ex. 23.

Suggestion. — Let $k = h = 1$, in the formula $\dfrac{s}{4} \cdot \dfrac{4k - 1}{2k + h}$.

10. If $AB = 12$, find the length of DO and the area of circle O.

11. If $AB = s$, find the length of DO and the area of circle O.

$$Ans. \ (1) \ \frac{s}{4}\sqrt{5}; \ (2) \ \frac{s^2\pi}{16}.$$

12. If the radius of circle O is 6 ft., find (1) the span AB; (2) the altitude DC; (3) the altitude of the arch AED.

$$Ans. \ (1) \ 24 \ ft.; \ (2) \ 12\sqrt{3}; \ (3) \ 6\sqrt{3}.$$

13. Find the lengths of the segments mentioned in Ex. 12 if the radius of circle O is r. $\qquad Ans.$ (1) $4r$; (2) $2r\sqrt{3}$; (3) $r\sqrt{3}$.

14. If the altitude of the arch AED is 4 ft., find (1) the span AB; (2) the altitude CD; (3) the radius of circle O.

$$Ans. \ \tfrac{16}{3}\sqrt{3}; \ (2) \ 8; \ (3) \ \tfrac{4}{3}\sqrt{3}.$$

15. If the altitude of the arch AED is r, find the lengths of the segments mentioned in Ex. 14.   *Ans.* (1) $\frac{4}{3}r\sqrt{3}$; (2) $2r$; (3) $\frac{r}{3}\sqrt{3}$.

16. If $AB = 12$, find the areas of the following figures:

(*a*) the curved $\triangle ACB$.   *Ans.* $12(4\pi - 3\sqrt{3})$.

Suggestion.—The sector CBA (B as center of circle) is $\frac{1}{6}$ of the area of the circle or 24π.   The area of the equilateral $\triangle ACB = 36\sqrt{3}$. The area of the segment bounded by the arc AC and the chord $AC = 24\pi - 36\sqrt{3}$.   The area of the curved $\triangle ACB$ is the area of the sector CAB (A as center) and the segment AC or $48\pi - 36\sqrt{3} = 12(4\pi - 3\sqrt{3})$.

(*b*) The curved triangle AED.   *Ans.* $3(4\pi - 3\sqrt{3})$.

17. Show that the area of the curved $\triangle ACB$ is equal to the area of a segment whose arc is 120°.

18. If $AB = s$, find the areas mentioned in Ex. 16.

$$Ans. \ \frac{s^2}{12}(4\pi - 3\sqrt{3}); \ (2) \ \frac{s^2}{48}(4\pi - 3\sqrt{3}).$$

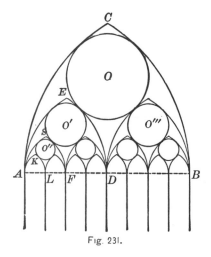

Fig. 231.

Fig. 231*a*. — From Lincoln Cathedral, England.

EXERCISES

277. 1. Construct Fig. 231.

Suggestion. — This figure involves Fig. 230.

2. If $AB = s$, find (1) the radii of circles O, O', and O'', and (2) the altitudes of arches ASF, AED, and ACB.

$$Ans. \text{ (1) } \frac{s}{4}, \frac{s}{8}, \frac{s}{16}; \text{ (2) } \frac{s}{8}\sqrt{3}, \frac{s}{4}\sqrt{3}, \frac{s}{2}\sqrt{3}.$$

3. If the radius of circle O is 5 ft., find (1) the radii of circles O' and O'', and (2) the altitudes of arches ASF, AED, and ACB.

$$Ans. \text{ (1) } 2\tfrac{1}{2} \text{ ft., } 1\tfrac{1}{4} \text{ ft ; (2) } 2\tfrac{1}{2}\sqrt{3}, 5\sqrt{3}, 10\sqrt{3}.$$

4. If the altitude of arch ACB is 12 ft., find (1) the span AB and (2) the radii of circles O, O', and O''.

$$Ans. \text{ (1) } 8\sqrt{3}; \text{ (2) } 2\sqrt{3}, \sqrt{3}, \tfrac{1}{2}\sqrt{3}.$$

278. Subtype. — Figures 232–234 show a subtype which may be explained by reference to Fig. 229, (see § 275, Ex. 23 and 25), or may give rise to the formulæ obtained below.

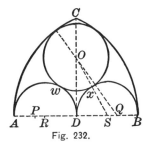

Fig. 232.

Fig. 232*a*. — From Amiens Cathedral.

In Fig. 232 $AQ = BP = \frac{5}{6} AB$. P and Q are the centers for the arcs CB and CA, respectively. $DA = DB$. Semicircles are constructed on AD and DB.

EXERCISES

1. Find the radius of circle O tangent to the arcs AC, CB, AwD, and DxB, if $AB = s$.

$$Ans. \frac{7s}{26}.$$

2. Construct circle O.

3. If $AQ = kAB = ks$, find the radius of circle O.

$$Ans. \frac{s}{2} \cdot \frac{4k-1}{4k+1}.$$

4. Show that the radius of circle O may be derived from the formulae obtained in §275, Ex. 23.

5. If $AQ = AB = s$, find the radius of circle O.   *Ans.* $\dfrac{3\,s}{10}$.

Suggestion. — Substitute $k = 1$ in the result obtained in Ex. 3.

6. Draw a figure to illustrate the case $k = 1$.

Suggestion. — See Fig. 232 b.

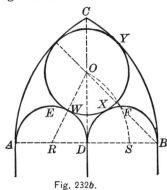

Fig. 232b.

7. In the figure obtained in Ex. 6, what per cent of the total area above the line AB is occupied by the included circle and semicircles?

Ans. 78% about.

8. In Fig. 232 if $AQ = \frac{1}{2}\,AB = \frac{1}{2}\,s$, find the radius of circle O and draw an illustrative figure.   *Ans.* $\dfrac{s}{6}$.

Suggestion. — See Fig. 234.

9. If $AQ = ks$, what value of k will make the radius of circle O equal the radius of one of the semicircles?   *Ans.* $k = \frac{3}{4}$.

Suggestion. — Solve the equation $\dfrac{s(4\,k - 1)}{8\,k + 2} = \dfrac{s}{4}$.

10. Verify the result obtained in Ex. 9, by a figure.

11. Construct a figure in which the arch ACB is stilted by $\frac{1}{6}$ the span and $AQ = \frac{5}{6}\,AB$. Find the radius of circle O in this figure.

12. Show that if the arch ACB is stilted by an amount equal to n, and if $AQ = ks$, the radius of circle O may be obtained from equation,

$$\sqrt{\left(ks - x\right)^2 - \left(ks - \frac{s}{2}\right)^2} = \sqrt{\left(\frac{s}{4} + x\right)^2 - \left(\frac{s}{4}\right)^2} - n.$$

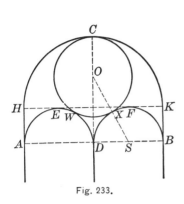

Fig. 233.

Fig. 233a. — From North Chicago Hebrew Congregation, Chicago, Ill.

279. In Fig. 233 HK is the diameter of the semicircle HCK. AB is parallel to HK. HA and KB are straight lines perpendicular to HK and equal to ¼ HK. The semicircles AED and DFB are constructed on ½ AB as diameter.

EXERCISES

1. Find the radius of circle O tangent to the three semicircles HCK, AED, and DFB, if $AB = s$. $Ans.\ \dfrac{9\,s}{32}.$

Suggestion. — Let the radius of the circle O be r. $\triangle ODS$ gives the equation $$\left(\frac{s}{4} + r\right)^2 = \left(\frac{s}{4}\right)^2 + \left(\frac{3\,s}{4} - r\right)^2.$$

2. Construct circle O.

3. What per cent of the figure above the line AB is occupied by the included circle and semicircle? *Ans.* about 69 %.

4. What loci would be needed to find the centers of the two small circles shown in the photograph?

Suggestion. — The locus needed is the locus of centers of circles tangent to each of two tangent circles. Its construction is beyond the province of elementary geometry.

5. If $HA = ks$, find the radius of circle O.

$$Ans.\ \frac{s}{2} \cdot \frac{(1 + 2\,k)^2}{3 + 4\,k}.$$

Suggestion. — △ *ODS* gives the equation

$$\left(\frac{s}{2}(1 + 2\,k) - r\right)^2 = \left(\frac{s}{4} + r\right)^2 - \left(\frac{s}{4}\right)^2.$$

6. Find the radius of circle *O* if *HA* is zero.   *Ans.* $\dfrac{s}{6}$.

Suggestion.,— Substitute $k = 0$ in the result obtained in Ex. 5.

7. Draw a figure to illustrate Ex. 6.   See Fig. 234.

8. If $HA = ks$, what value of k will make the radius of circle *O* equal to the radius of the semicircle *AED*?

$$Ans.\ k = \frac{\sqrt{3} - 1}{4}.$$

Suggestion. — Solve the equation, $\dfrac{s}{2} \cdot \dfrac{(1 + 2\,k)^2}{3 + 4\,k} = \dfrac{s}{4}.$

280. In Fig. 234 AD=DB and the three semicircles ACB, AED, and DFB are drawn on AB, AD, and DB, respectively, as diameters.   The circles O, O′, and O″ are tangent as indicated.

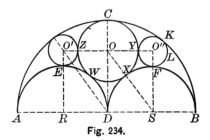

Fig. 234.

EXERCISES

1. Find the radius of circle *O* tangent to each of the three semicircles, if $AB = s$.   *Ans.* $\dfrac{s}{6}$.

Suggestion. — Let r be the radius of circle *O*. △ *ODS* gives the equation, $\left(\dfrac{s}{2} - r\right)^2 + \left(\dfrac{s}{4}\right)^2 = \left(\dfrac{s}{4} + r\right)^2.$

2. Show how to construct circle *O* as indicated.

3. Erect $O'R$ perpendicular to AD at R the middle point of AD. Find the radius of circle O' tangent to semicircle AED at E and to semicircle ACB, if $AB = s$.   *Ans.* $\dfrac{s}{12}$.

Suggestion. — Let the radius of circle O' be r'. △ $O'RD$ gives the equation $\left(\dfrac{s}{4} + r'\right)^2 + \left(\dfrac{s}{4}\right)^2 = \left(\dfrac{s}{2} - r'\right)^2.$

4. Prove that this circle will be tangent to circle O.

Suggestion. — $O'R$ is parallel to OD. $O'R = \dfrac{4RD}{3} = OD$. Hence

$O'O$ is parallel and equal to RD. But $r + r' = \dfrac{s}{12} + \dfrac{s}{6} = \dfrac{s}{4} = RD$.

Therefore the line of centers of circles O and O' equals the sum of the radii, and the circles are tangent.

5. Show how to construct circle O'.[1]

Suggestion. — Since OO' is parallel to RD, perpendiculars erected at R to line AD and at O to line CD intersect at O'.

6. What fraction of the semicircle ACB is occupied by the included circles and semicircles? *Ans.* $\frac{5}{6}$.

281. Type II. — Figures 196, 235–241, 245, 250–252, and 257 furnish illustrations of this type, the formula for which is obtained from Ex. 3 below.

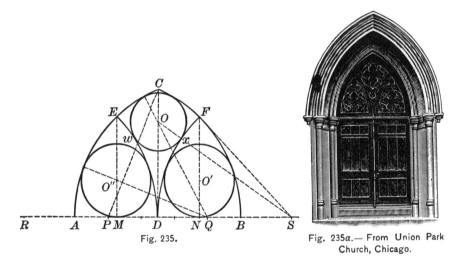

Fig. 235.

Fig. 235a.— From Union Park Church, Chicago.

In Fig. 235 $AQ = PB = \frac{4}{5} AB$. P and Q are centers for the arcs CB and AC, respectively. Arcs ED and FD are drawn with R and S as centers and AQ as radius.

[1] Hanstein, Pl. 19.

EXERCISES

1. Find the radius of circle O inscribed in the curved quadrilateral $CEDF$, if $AB = s$. *Ans.* $\dfrac{11\,s}{64}$.

Suggestion. — Let r be the radius of circle O. Solve for r in the equation, $\left(\dfrac{4\,s}{5} - r\right)^2 - \left(\dfrac{3\,s}{10}\right)^2 = \left(\dfrac{4\,s}{5} + r\right)^2 - \left(\dfrac{4\,s}{5}\right)^2$.

2. If $AB = s$, find the lengths of OD and FN.

Ans. (1) $\dfrac{9\,s}{320}\sqrt{385}$; (2) $\dfrac{3\,s}{20}\sqrt{15}$.

3. If $AQ = kAB = ks$, find the radius of circle O.

Ans. $\dfrac{s}{16}\left(\dfrac{4\,k-1}{k}\right)$.

Suggestion. — $\triangle\, ODQ$ and ODS give the equation

$$(ks - r)^2 - \left(ks - \dfrac{s}{2}\right)^2 = (ks + r)^2 - k^2 s^2.$$

4. If $AQ = \frac{5}{6}\,AB = \dfrac{5\,s}{6}$, find the radius of circle O by substituting $k = \frac{5}{6}$ in the result obtained in Ex. 3. *Ans.* $\dfrac{7\,s}{40}$.

Draw a figure to illustrate this case. Let $AB = 5$ in.

5. If $AQ = AB = s$, that is, if $k = 1$, find the radius of circle O, and draw an illustrative figure. *Ans.* $\dfrac{3\,s}{16}$.

6. If $AQ = \frac{1}{2}\,AB = \dfrac{s}{2}$, find the radius of circle O, and draw an illustrative figure. (See Fig. 240.) *Ans.* $\dfrac{s}{8}$.

7. Give construction for circle O''.

Suggestion. — See § 242, Ex. 1.

8. If $AQ = \frac{4}{5}\,AB = \frac{4}{5}\,s$, find the radius of circle O''. *Ans.* $\dfrac{27\,s}{128}$.

Suggestion. — Join O'' with the center for the arc EA. Use $\triangle\, O''QM$.

9. If $AQ = kAB = ks$, find the radius of circle O''.

Ans. $\dfrac{s}{32} \cdot \dfrac{8\,k-1}{k}$.

10. Construct the entire figure for the case $k = 1$.

Suggestion. — By substituting in the necessary formulæ, the radii of circles O and O'' are $\dfrac{3\,s}{16}$ and $\dfrac{7\,s}{32}$, respectively.

11. If $AQ = kAB$, find the ratio between the radii of circles O and O''. $\qquad Ans.\ \dfrac{8\,k-2}{8\,k-1}.$

12. Show that if $AQ = kAB$, it is impossible to find a value for k that will make circle O equal to circle O''.

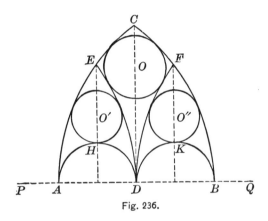

Fig. 236.

EXERCISES

282. **1.** Give the construction for Fig. 236. The radius for the arcs CA and CB is $\frac{5}{4}\,AB$.

Suggestion. — For the construction of circles O' and O'', see § 273, Ex. 6.

2. Find the radius of circles O and O', if the radius for the arcs AC and CB is $kAB = ks$. $\qquad Ans.\ (2)\ \dfrac{s}{4}\cdot\dfrac{4\,k-1}{4\,k+1}.$

3. Show that the result obtained in Ex. 2 may be derived from that in § 273, Ex. 6.

4. Construct the entire figure if the arch ACB is equilateral. Let $AB = 5$ in.

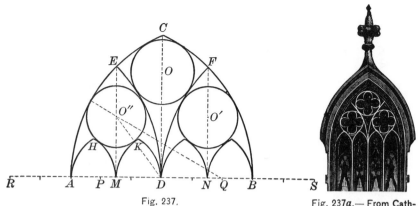

Fig. 237.

Fig. 237a.— From Catholic Church, Evansville, Indiana.

283. In Fig. 237 $AQ = BP = \frac{5}{6} AB$. P and Q are the centers for the arcs CB and AC, respectively. Arcs ED and DF are drawn with AQ as radius and R and S as centers. Arches AHM, MKD, etc. are equilateral. $AM = MD = DN = NB$.

EXERCISES

1. Find the radius of circle O'' tangent to the arcs EA, ED, HM, and KM, if $AB = s$. *Ans.* $\dfrac{17\,s}{104}$.

2. If $AQ = kAB = ks$, find the radius of circle O''.

Ans. $\dfrac{s}{8} \cdot \dfrac{8k - 1}{4k + 1}$.

3. Show that the formula obtained in Ex. 2 may be derived from that in § 275, Ex. 23.

4. Construct the entire figure if $AQ = AB = 5$ in.

Suggestion. — For the radius of circle O see § 281, Ex. 3.

5. If $AQ = kAB$, for what value of k will the radius of circle O equal the radius of circle O'? *Ans.* $k = \frac{1}{2}$.

Suggestion. — Solve the equation $\dfrac{4k - 1}{16\,k} = \dfrac{8k - 1}{8(4k + 1)}$.

6. Make a drawing to illustrate the case $k = \frac{1}{2}$.

7. Show how to construct the quadrifoils shown in the circles in Fig. 237 a.

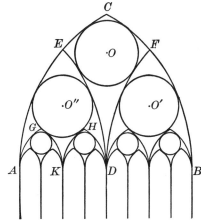

284. In Fig. 238, arch ACB is equilateral. Arcs ED and FD are drawn with AB as radius, and centers on AB extended. Arches AGK, KHD, etc., are equilateral.

Fig. 238.—From Choir, Tintern Abbey, England. Bond, p. 475.

EXERCISES

1. If $AB = s$, find the radius of circle O and circle O'.

$$Ans. \quad (1) \; \frac{3s}{16}; \quad (2) \; \frac{7s}{40}.$$

2. Construct the entire figure.

3. Show that the radius of circle O' may be derived from the formula obtained in § 275, Ex. 23.

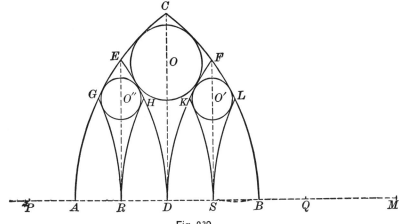

Fig. 239.

Fig. 239a. — From Wells Cathedral Chapter House.

285. In Fig. 239 AQ=BP= $\frac{5}{4}$ AB. P and Q are the centers for the arcs BC and AC, respectively. AR=RD= DS=SB. Arcs GR, ED, KS, etc., are drawn with AQ as radius, and points on line PM as centers. M is the center for the arc FD.

EXERCISES

1. Find the radius of circle O inscribed in the quadrilateral $CEDF$.

Suggestion. — See § 281, Ex. 3.   *Ans.* $\frac{s}{5}$.

2. If $AQ = \frac{5}{4} AB = \frac{5}{4} s$, find the radius of circle O'.   *Ans.* $\frac{9 s}{80}$.

3. If $AQ = ks$, find the radius of circle O' where $AB = s$.

Ans. $\frac{s}{64} \cdot \frac{8 k - 1}{k}$.

4. If $AQ = \frac{5}{4} s$, find the radius of circle O' from the formula obtained in § 281, Ex. 3.

Suggestion. — From § 281, Ex. 3, $r = \dfrac{DB}{16} \cdot \dfrac{4 k - 1}{k}$, where k is the relation between DM and DB.   Hence $k = \dfrac{5}{2}$, $DB = \dfrac{s}{2}$.   Make the necessary substitutions.

5. Show that the formula obtained in Ex. 3 may be derived from that in § 281, Ex. 3.

Suggestion. — Since $\dfrac{DM}{DB} = \dfrac{2\,DM}{AB}$ change k to $2\,k$.   Since $DB = \dfrac{1}{2} AB$, change s to $\dfrac{s}{2}$ in formula obtained in § 281, Ex. 3.

6. If $AQ = \frac{4}{5} AB = \frac{4}{5} s$, find the radius of circle O and of circle O' and draw an illustrative figure.   Let $AB = 8$ in.

7. If $AQ = \frac{3}{4} AB = \frac{3}{4} s$, find the radius of circle O and of circle O' and draw an illustrative figure. $AB = 6$ in. *Ans.* (1) $\frac{s}{6}$; (2) $\frac{5s}{48}$.

8. If $AQ = AB = s$, find the radius of circle O and of circle O' and draw an illustrative figure. *Ans.* (1) $\frac{3s}{16}$; (2) $\frac{7s}{64}$.

9. If $AQ = kAB = ks$, for what value of k will circles O and O' have the same radius? *Ans.* $k = \frac{3}{8}$.

Suggestion. — Solve the equation $\dfrac{s(8k-1)}{64k} = \dfrac{s(4k-1)}{16k}$.

10. Draw a figure to illustrate the case $k = \frac{3}{8}$. (See Fig. 239 *b.*)

11. If $AQ = {}^{3}AB$ and the radius of circle O' is 18, find AB, and the altitudes CD and FS.

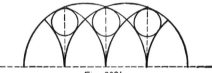

Fig. 239*b*.

Ans. (1) $172\frac{4}{5}$; (2) $\dfrac{432\sqrt{2}}{5}$; (3) $\dfrac{216}{5}\sqrt{5}$.

12. If $AQ = {}_{\overline{6}} AB$ and the radius of circle O is 2, find span AB and the altitudes CD and FS. *Ans.* (1) $11\frac{3}{7}$; (2) $\frac{40}{21}\sqrt{21}$; (3) $\frac{20}{21}\sqrt{51}$.

286. In Fig. 240 AB is the diameter of the semicircle ACB. A and B are the centers and $\frac{1}{2}$ AB the radius for the arcs DE and DF. Circle O is tangent to the arcs as shown.

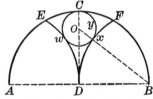

Fig. 240. — From Great Waltham Church, Essex, Eng. Brandon (I), Sect. II, Woodwork, Pl. 2.

EXERCISES

1. Prove that the points C, O, and D are collinear.

2. Prove that the arcs ED and FD are tangent at point D.

3. If circle O is tangent to the arc DF at point x, prove that the points O, x, and B are collinear.

4. If O is the center of the circle, prove that OD is perpendicular to AB.

5. If $AB = s$, find the radius of circle O.   *Ans.* $\dfrac{s}{8}$.

Suggestion. — Let the radius of circle O be r. $\triangle ODB$ gives the equation, $\left(\dfrac{s}{2}\right)^2 + \left(\dfrac{s}{2} - r\right)^2 = \left(\dfrac{s}{2} + r\right)^2$.

6. Show how to construct the entire figure.

7. Show that Fig. 240 is a special case of Fig. 235.

8. If $AB = s$, find the area of the curved figures: (*a*) triangle AED; (*b*) triangle EFD.

Ans. (*a*) $\dfrac{s^2}{48}(4\pi - 3\sqrt{3})$; (*b*) $\dfrac{s^2}{24}(3\sqrt{3} - \pi)$.

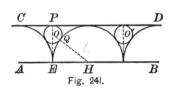

Fig. 24I.

Fig. 24Ia. — Ospedale Maggiore, Milan.

287. In Fig. 241 a series of semicircles is inscribed between the two parallel lines **AB** and **CD** as shown, and circle **O** is tangent to arcs **CE** and **EQ**.

EXERCISES

1. Prove that the center of circle O must lie on a perpendicular to AB at E.

Suggestion. —Use the theorem : " The common tangent to two equal tangent circles is a part of the locus of centers of circles tangent to the two given circles."

2. If circle O and arc EQ are tangent at Q, prove that the points H, Q and O are collinear.

3. Find the radius of circle O tangent to the arcs CE and EQ and to the line CD, if $EP = s$.   *Ans.* $\dfrac{s}{4}$.

Suggestion. — In $\triangle EHO$, $\overline{OH}^2 = \overline{OE}^2 + \overline{EH}^2$. Let r be the radius of circle O, and $s = EH = EP$. Since $OH = s + r$ and $OE = s - r$, the equation becomes $(s + r)^2 = (s - r)^2 + s^2$.

4. Show how to construct the entire figure.

5. If $EP = s$, find the area of the curved figure of which PCE is half. $\qquad\qquad$ *Ans.* $\dfrac{s^2}{2}(4 - \pi)$.

6. Can the answer to Ex. 3 be obtained from that of § 286, Ex. 5?

TYPES OF THREE-LIGHT WINDOWS

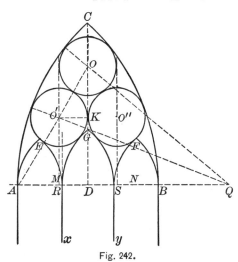

Fig. 242.

Fig. 242a. — From Lincoln Cathedral, England.

288. Type I. — In Fig. 242 $AQ = \tfrac{3}{2}$ AB. AQ is the radius for the arcs CB and CA. Arches AER and SFB are equilateral and AR = RS = SB.

EXERCISES

1. Find the radius of circle O' tangent to the arcs CA and ER and the line CD. $\qquad\qquad$ *Ans.* $\dfrac{5\,s}{24}$.

Suggestion. — Draw $O'Q$ and $O'A$. In $\triangle AO'Q$, $\overline{O'A}^2 - \overline{AM}^2 =$ $\overline{O'Q}^2 - \overline{MQ}^2$ or $\left(\dfrac{s}{3} + r\right)^2 - \left(\dfrac{s}{2} - r\right)^2 = \left(\dfrac{3}{2}s - r\right)^2 - (s + r)^2$. Why does $MD = r$?

2. Construct circles O'' and O'. Prove that they are tangent to each other.

3. If $AB = s$ and $AQ = \dfrac{3\,s}{2}$, find the length of $O'M$.

$$Ans. \quad \frac{s}{12}\sqrt{30}.$$

4. Construct arcs GR and GS tangent to circles O' and O'', respectively, and to lines Rx and Sy at R and S respectively.

Suggestion.—This involves the problem, " To construct a circle tangent to a given line at a given point and to a given circle." See § 246. Point N is the center for the arc RG.

5. Find the radius of circle O tangent to arcs CA and CB and to circles O'' and O', if $AB = s$.   *Ans.* .208 s.

Suggestion. — $OD = KD + KO$, $KO = \sqrt{\left(\dfrac{5\,s}{24} + x\right)^2 - \left(\dfrac{5\,s}{24}\right)^2}$.

$DO = \sqrt{\overline{OQ}^2 - \overline{DQ}^2}$.   Hence get the equation,

$$\sqrt{\left(\frac{3\,s}{2} - x\right)^2 - s^2} = \frac{s}{12}\sqrt{30} + \sqrt{\frac{5\,sx}{12} + x^2}.$$

6. What is the per cent of error if the radius of circle O is said to be equal to the radius of circle O'' ?

7. If $AQ = kAB = ks$, find the radius of circle O'.

$$Ans. \quad \frac{s}{6} \cdot \frac{9\,k - 1}{6\,k + 1}.$$

Suggestion. — Get the equation,

$$(ks - x)^2 - \left(ks - \frac{s}{2} + x\right)^2 = \left(\frac{s}{3} + x\right)^2 - \left(\frac{s}{2} - x\right)^2.$$

8. If $AQ = kAB$, for what value of k will the radius of circle O' be $\dfrac{AB}{6}$?   *Ans.* $k = \frac{2}{3}$.

Suggestion. — Solve the equation $\dfrac{s}{6} \cdot \dfrac{9\,k - 1}{6\,k + 1} = \dfrac{s}{6}$.

9. If $AQ = kAB = ks$, find the radius of circle O.

Suggestion. — This involves the equation,

$$\sqrt{(ks - r)^2 - \left(ks - \frac{s}{2}\right)^2} = \sqrt{\left(\frac{s}{3} + p\right)^2 - \left(\frac{s}{2} - p\right)^2} + \sqrt{(p + r)^2 - p^2},$$

where p is the radius of circle O' or $\dfrac{s}{6} \cdot \dfrac{9\,k - 1}{6\,k + 1}$.

10. If $AQ = kAB = ks$ and the radius of the arcs AE, ER, SF, and FB is $\dfrac{ks}{3}$, find the radius of circle O'. $Ans.$ $\dfrac{s}{42}(9 - k)$.

$Suggestion.$ — Get the equation

$$(ks - r)^2 - \left(ks - \frac{s}{2} + r\right)^2 = \left(\frac{ks}{3} + r\right)^2 - \left(\frac{s}{2} - r\right)^2.$$

11. If $AQ = kAB = ks$, and the radius of the arcs AE, ER, SF, and FB is $\dfrac{ks}{3}$, find the radius of circle O.

289. Type II. Figures 243–247 and 260 furnish illustrations of this type, the formula for which is obtained in Ex. 4 below.

In Fig. 243 $AQ = BP = \tfrac{5}{6}AB$, P and Q are the centers for the arcs CB and AC, respectively. $AR = RS = SB$. $RN = BN' = \tfrac{5}{6}RB$. N and N' are centers for the arcs RF and BF. Arch AES is similarly drawn on AS.

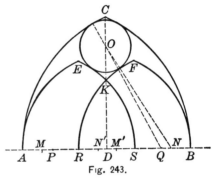

Fig. 243.

EXERCISES

1. Find the radius of circle O tangent to the arcs CA, CB, ES, and RF, if $AB = s$. $Ans.$ $\dfrac{23s}{150}$.

$Suggestion.$ — $\triangle\,OQD$ and OND give the equation,

$$\left(\frac{5s}{6} - r\right)^2 - \left(\frac{s}{3}\right)^2 = \left(\frac{5s}{9} + r\right)^2 - \left(\frac{7s}{18}\right)^2.$$

2. If $AQ = kAB = ks$ and $RN = SM = \tfrac{2}{3}ks$, find the radius of circle O. $Ans.$ $\dfrac{s}{30}\left(\dfrac{7k - 2}{k}\right)$.

$Suggestion.$ — Solve for r in the equation,

$$(ks - r)^2 - \frac{s^2}{4}(2k - 1)^2 = \left(\frac{2ks}{3} + r\right)^2 - \frac{s^2}{36}(4k - 1)^2.$$

3. If $k = 1$, find the radius of circle O by substituting in the result obtained in Ex. 2.   *Ans.* $\dfrac{s}{6}$.

Note.— This is Fig. 244.

4. If $AQ = ks$ and $RN = h \cdot \dfrac{2\,s}{3}$ find the radius of circle O.

$$Ans. \quad \frac{s}{2} \cdot \left(\frac{9\,k - 2\,h - 2}{9\,k + 6\,h} \right).$$

5. Show how the formula in **Ex. 2** may be obtained from that in **Ex. 4.**

290. Figure 244 is a special case under Fig. 243. It is readily solved by the special method indicated below, or by the formula in § 289, Ex. 2.

In Fig. 244 the arch ACB is equilateral. AR = RS = SB. Arches AES and RFB are equilateral on AS and RB as spans.

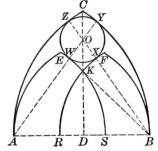

Fig. 244. — From Church in Hertfordshire, England.   Brandon (I), p. 28.

EXERCISES

1. If D is the center of AB, prove that the points C, K, and D are collinear.

Suggestion. — Use the theorem, "If two equal circles intersect, the common chord is the perpendicular bisector of the line of centers."

2. If circle O is tangent to the arcs AC and BC or to the arcs ES and RF, prove that its center must lie on line CD.

3. If circle O is tangent to the concentric arcs AC and FR, where must its center lie?

4. If circle O is tangent to the concentric arcs BC and ES, where must its center lie?

5. Show that the radius of circle O is $\dfrac{AB}{6}$, and construct circle O.

Suggestion. — It is necessary to prove that $OX = OY = OZ = OW$, and that circle O is tangent to the four arcs AC, BC, ES, and FR.

6. Show that the radius of circle O may be obtained from the formula found in § 289, Ex. 2.

7. If $AB = s$, find the length of the line OD, the altitude of the arch AES, and the length of KD.

$$Ans. \ (1) \ \frac{2s}{3}; \ (2) \ \frac{s}{3}\sqrt{3}; \ (3) \ \frac{s}{6}\sqrt{7}.$$

291. Figure 245 shows a six-light window which is a combination of the three-light type shown in Fig. 243, and the two-light type shown in Fig. 235.

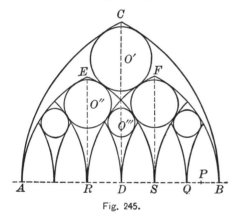

Fig. 245.

Fig. 245a. — From the West Front, Lichfield Cathedral, England.

In Fig. 245 AQ is the radius of $\overset{\frown}{AC}$ and $\overset{\frown}{BC}$. $AQ = \frac{5}{6}$ AB. $AR = RS = SB$. PR is the radius of $\overset{\frown}{EA}$, $\overset{\frown}{ES}$, $\overset{\frown}{FR}$, and $\overset{\frown}{FB}$. $PR = \frac{7}{8}$ RB. All of the smaller arcs are drawn with PR as radius.

EXERCISES

1. Find the radius of circle O'. $Ans. \ \dfrac{5s}{34}.$

2. Show that the answer to Ex. 1 may be derived from the formula in § 289, Ex. 4.

3. Find the radius of circle O''. $Ans. \dfrac{5\,s}{42}$

4. Show that the answer to Ex. 3 may be derived from the formula in § 281, Ex. 3.

5. Find the radius of circle O'''. $Ans. \dfrac{s}{14}$.

6. Show that the answer to Ex. 5 may be derived from the formula in § 281, Ex. 3.

292. Figure 246 shows a **sub-type** which may be explained by reference to Fig. 243 (see § 289, Ex. 4), or may give rise to the formulæ obtained below.

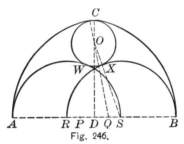

Fig. 246.

Fig. 246a. — From Masonic Hall, 91st St., Chicago, Ill.

In Fig. 246 AQ = BP = $\frac{3}{5}$ AB. P and Q are the centers for the arcs CB and AC, respectively. AR = RS = SB. Semicircles AWS and RXB are drawn on AS and RB as diameters.

EXERCISES

1. Find the radius of circle O tangent to the arcs AC, CB, WS, and RX, if $AB = s$. $Ans. \dfrac{s}{7}$.

2. Show how to construct the entire figure.

3. If $AQ = kAB = ks$, find the radius of circle O.

$$Ans. \ \frac{s}{2} \cdot \frac{3\,k-1}{3\,k+1}.$$

4. Show that the formula obtained in Ex. 3 may be derived from that in § 289, Ex. 4.

5. If $AQ = AB$, find the radius of circle O. $Ans. \dfrac{s}{4}$.

Suggestion. — Substitute $k = 1$ in the result obtained in Ex. 3.

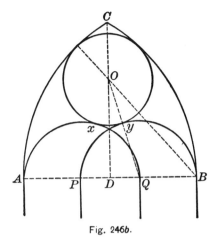

Fig. 246b.

6. Draw a figure to illustrate Ex. 5. (See Fig. 246 a.)

7. In the figure obtained in Ex. 6, what per cent of the entire area is not occupied by the included circle and semicircle?

Ans. about 22 %.

8. In Fig. 246, if $AQ = kAB = ks$, what value of k will make the radius of circle O equal the radius of the semicircle RXB? Draw an illustrative figure. *Ans.* $k = \frac{5}{8}$.

Suggestion. — To obtain the value of k, solve the equation,

$$\frac{s}{2} \cdot \frac{3k - 1}{3k + 1} = \frac{s}{3}.$$

9. What value of k will make the diameter of circle O equal to the line RS? *Ans.* $k = \frac{2}{3}$.

10. If $AQ = \frac{1}{3} AB$, find the radius of circle O and draw an illustrative figure. *Ans.* $\frac{s}{10}$.

11. Find the radius of circle O if $AQ = kAB$ and $AS = mAB$, where $AB = s$. *Ans.* $\frac{s}{2}\left(\frac{2k - m}{2k + m}\right)$.

12. Draw a figure to illustrate the case $k = \frac{1}{2}$, $m = \frac{3}{4}$, and find the radius of circle O for this case. *Ans.* $\frac{s}{14}$.

13. Show that the formula obtained in Ex. 3 may be derived from that in Ex. 11.

14. Can the formula obtained in Ex. 11 be derived from that in § 289, Ex. 4?

15. Show how to construct the trefoil shown in the circle in Fig. 246a.

293. Figure 247 shows a window which is a combination of that suggested in § 292, Ex. 12, with the one-light type shown in Fig. 228 and with the two-light type shown in Fig. 235.

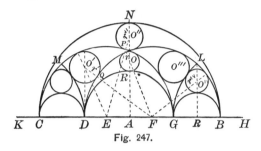

Fig. 247.

Fig. 247a. — From Or San Michele, Florence, Italy.

In Fig. 247 $\overgroup{CNB}$ is a semicircle on CB as diameter. CD = DA = AG = GB. Semicircles are constructed on the line segments CG, DB, CD, DG, and GB as diameters. The circles are tangent to the arcs as shown. Arcs DM and LG are drawn with DF as radius and centers on the line KH.

EXERCISES

1. Find the radii of circles O and O^{iv}. Show that they may be derived from the formula obtained in § 273, Ex. 6.

$$Ans. \ (1) \ \frac{s}{20}; \ (2) \ \frac{s}{16}.$$

2. Find the radius of circle O''. Show that it may be derived from the formula in § 292, Ex. 11.

$$Ans. \ \frac{s}{14}.$$

3. Find the radius of circle O'. Show that it may be derived from the formula in § 281, Ex. 3.

$$Ans. \ \frac{s}{12}.$$

4. Construct the entire figure, using $AB = 7$ in.

5. What per cent of the total figure is occupied by the six circles?

6. Construct a figure similar to Fig. 247, using instead of the semicircles on CB, CG, and DB pointed arches with the radius for the arcs $\frac{3}{4}$ of the span.

294. Type III. Figures 248–252 and 257–261 furnish illustrations of this type. The formula to which they may be referred is obtained in Ex. 3 below.

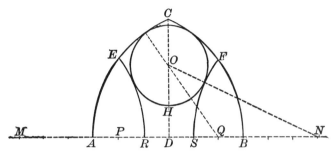

Fig. 248. — From Merton College Chapel, Oxford, Eng. A. and A. W. Pugin, Vol. I, Pl. 5.

In Fig. 248 $AQ = BP = \frac{5}{6} AB$. P and Q are the centers for the arcs CB and AC, respectively. $AR = RS = SB$. Arcs ER and FS are drawn with M and N as centers and AQ as radius.

EXERCISES

1. Find the radius of circle O tangent to the arcs AC, CB, ER, and FS, if $AB = s$. *Ans.* $\dfrac{4\,s}{15}$.

2. Construct the entire figure, using $AB = 5$ in.

3. If $AQ = kAB = ks$ and $AR = mAB = ms$, find the radius of circle O. *Ans.* $\dfrac{s}{4} \cdot \dfrac{(2\,k - m)(1 - m)}{k}$.

Suggestion. — The equation is

$$(ks - r)^2 - \frac{s^2}{4}(2\,k - 1)^2 = (ks + r)^2 - \frac{s^2}{4}(2\,k + 1 - 2\,m).$$

4. Let $k = 1$ and $m = \frac{1}{3}$. Find the radius of the circle by substituting in the formula obtained in Ex. 3. Draw an illustrative figure.

Ans. $\dfrac{5\,s}{18}$.

5. If $k = 1$ and $m = \frac{1}{2}$, find the radius of the circle by substituting in the formula obtained in Ex. 3. Draw an illustrative figure. Compare with Fig. 235. Compare the result obtained with that in § 281, Ex. 5.

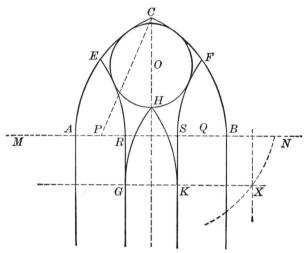

Fig. 249. — From Hyde Park Presbyterian Church, Chica o, Ill.

295. Figure 249 is identical with Fig. 248 with the addition of the arch GHK. The arcs GH and HK are constructed with AQ as radius, tangent to lines RG and SK, respectively, and passing through point H.

EXERCISES

1. Show how to construct the arcs *GH* and *HK*.

Suggestion.— This involves the construction of a circle by intersecting loci. What loci are required?

2. Show that in general two circles can be constructed to fit the given conditions. (See Fig. 249a.)

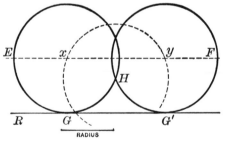

Fig. 249a.

3. Are the arches *GHK* and *SFB* congruent if *GK* is made to coincide with *SB*?

4. Construct a figure similar to Fig. 249 with the arch *ACB* equilateral.

5. Construct a figure similar to Fig. 249, with $AQ = \frac{6}{5} AB$.

296. Figure 250 is a special case under the types shown in Figs. 248 and 235.

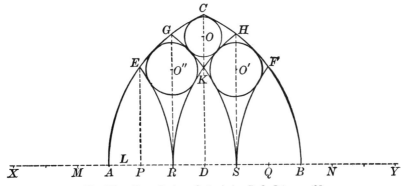

Fig. 250. — From Durham Cathedral. E. S. Prior, p. 195.

In Fig. 250 P and Q are the centers for the arcs CB and CA, respectively. $AQ = BP = \frac{5}{6} AB$. $AR = RS = SB$. The arcs ER, GS, HR, and FS are drawn with radius AQ and centers on line AB.

EXERCISES

1. Find radius of circle O inscribed in quadrilateral $CGKH$, if $AB = s$. *Ans.* $\dfrac{s}{10}$.

2. Verify the result obtained in Ex. 1 by substituting $k = \frac{5}{6}$, $m = \frac{2}{3}$ in the result shown in § 294, Ex. 3.

3. Show how to construct circle O.

4. If $AQ = BP = kAB = ks$, find the radius of circle O.

Ans. $\dfrac{s}{18} \cdot \dfrac{3k-1}{k}$.

5. Verify the result shown in Ex. 4 by substituting $m = \frac{2}{3}$ in the result obtained in § 294, Ex. 3.

6. If $AQ = BP = \frac{5}{6} AB = \frac{5}{6} s$, find the radius of circle O''.

Ans. $\dfrac{2s}{15}$.

7. Construct the entire figure when $AQ = \frac{5}{6} AB$. Let $AB = 5$ in.

8. If $AQ = kAB = ks$, find the radius of circle O''.

Ans. $\dfrac{s}{36} \cdot \dfrac{6k-1}{k}$.

9. Show that the formula found in Ex. 8 can be obtained from that in § 281, Ex. 3.

Suggestion. — From § 281, Ex. 3, $r = \dfrac{AS}{16} \cdot \dfrac{4\,k' - 1}{k'}$, where $k' = \dfrac{XR}{AS}$.

Since $\dfrac{XR}{AS} = \dfrac{3}{2}\dfrac{XR}{AB}$, $k' = \dfrac{3}{2}k$, $AS = \dfrac{2}{3}s$. Make the necessary substitutions.

10. Show that the formula found in Ex. 8 can be obtained from that in § 294, Ex. 3.

Suggestion. — From § 294, Ex. 3, $r = \dfrac{AS}{4}\dfrac{(2\,k' - m')(1 - m')}{k'}$, where m' and k' are the ratios $\dfrac{AR}{AS}$ and $\dfrac{XR}{AS}$, respectively. First substitute $m' = \tfrac{1}{2}$, and reduce to $\dfrac{AS}{16} \cdot \dfrac{4\,k' - 1}{k'}$

11. If $AQ = AB$, find the radius of circle O and of circle O' by substituting $k = 1$ in the results obtained in Exs. 4 and 8.

$$\textit{Ans.}\;\; (1)\;\frac{s}{9}\;;\;(2)\;\frac{5\,s}{36}.$$

12. Construct the entire figure when $k = 1$.

13. What value of k will make the radius of circle O one half the radius of circle O'? 　　　　　　　　　　　　*Ans.* $k = \tfrac{1}{2}$.

Suggestion. — Solve the equation $\dfrac{s}{36} \cdot \left(\dfrac{6\,k - 1}{k}\right) = \dfrac{2\,s}{18}\left(\dfrac{3\,k - 1}{k}\right)$.

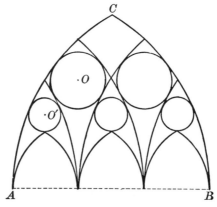

Fig. 25l. — From R. C. Church, Oys er Bay, N. Y.
See Arch. Rev., Vol. V, p. 76.

EXERCISES

297. **1.** If the arch ACB is equilateral, construct the diagram shown in Fig. 251.

Suggestion. — To find the center of circle O', use the problem, " To construct a circle tangent to a given circle and to a given line at a given point." See § 246.

2. Show that the radius of circle O may be derived from the formula in § 281, Ex. 3.

298. Figure 252 shows a four-light window which is a combination of the types discussed in §§ 281 and 294.

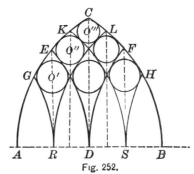

In Fig. 252 the arch ACB is equilateral. AR = RD = DS = SB. $\overset{\frown}{RG}$, $\overset{\frown}{ED}$, $\overset{\frown}{KS}$, etc., are drawn with radius AB and centers on AB extended.

Fig. 252.

EXERCISES

1. Find the radii of circles O'' and O''' if $AB = s$, and show that these may be obtained from the formula found in § 294, Ex. 3.

$$Ans. \ (1) \ \frac{3s}{32}; \ (2) \ \frac{5s}{64}.$$

2. Find the radius of circle O' and show that it may be obtained from the formula found in § 281, Ex. 3.

3. If $AB = s$, find the altitude of the arches RLB, DFB, and SHB.

4. What per cent of the entire figure is occupied by the six circles?

5. Construct the entire figure, using $AB = 8$ in.

TYPES OF FIVE-LIGHT WINDOWS

299. Type I.   Figures 253–256 furnish illustrations of this type, the formula for which is obtained in Ex. 5 below.   For other illustrations see Fig. 137a and the note under Ex. 4, § 300.

In Fig. 253 AQ = BP = $\frac{6}{7}$ AB. P and Q are the centers for the arcs CB and AC, respectively.   AM = NB = $\frac{2}{5}$ AB.   SN = RB = $\frac{6}{7}$ BN. R and S are the centers for the arcs FB and FN, respectively. Arch AEM is similarly drawn.

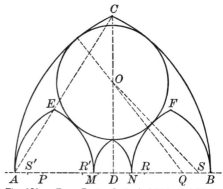

Fig. 253. — From Exeter Cathedral Aisle and Choir.
Sharpe (2).

EXERCISES

1. Find the radius of circle O tangent to the arcs AC, CB, EM, and FN. if $AB = s$.   *Ans.* $\dfrac{2s}{7}$.

2. If $AQ = kAB$ and $SN = kNB$, find the radius of circle O, if $AB = s$ and $NB = \frac{2}{5} s$.   *Ans.* $\dfrac{3s}{70} \dfrac{9k - 2}{k}$.

3. Prove $\triangle ACB$ and AEM similar and the points A, E, and C collinear.

Suggestion. — See § 275, Exs. 16–19.

4. If $AQ = AB$, find the radius of circle O.   *Ans.* $\dfrac{3s}{10}$.

5. If $AQ = kAB = ks$ and $SN = hNB = h \cdot \dfrac{2s}{5}$, find the radius of circle O.

$$Ans. \quad \frac{s}{10} \left(\frac{25k + 2h - 6}{5k + 2h} \right).$$

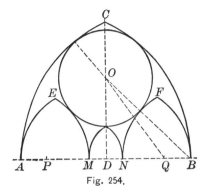

Fig. 254.

300. In Fig. 254 $AQ = BP = \frac{6}{7} AB$.
P and Q are the centers for the arcs CB
and AC, respectively. Arches AEM and
NFB are equilateral. $AM = NB = \frac{2}{5} AB$.

EXERCISES

1. Find the radius of circle O tangent to the arcs AC, CB, EM, and FN, if $AB = s$. *Ans.* $\dfrac{61\,s}{220}$.

2. If $AQ = kAB = ks$, and $BN = \frac{2}{5} AB$, find the radius of circle O.

$$Ans. \ \frac{s}{10} \cdot \frac{25\,k - 4}{5\,k + 2}.$$

3. Show that the results obtained in Ex. 1 and 2 may be derived from the formula in § 299, Ex. 5.

4. If $AQ = \frac{4}{5} AB$ and $BN = \frac{2}{5} AB$, find the radius of circle O.

Note. — The First Congregational Church, Portland, Oregon, contains a window of the proportions given in Ex. 4. The mullions in the arches AEM and NFB are of the type shown in Fig. 235.

5. Make a diagram for the window in the church referred to in Ex. 4, note.

6. Show how to construct the small arch erected on the span MN, if the arcs that form it are tangent to the arcs EM and FN at M and N respectively, and if its apex just touches the circle O.

7. What locus is needed in the construction called for in Ex. 6?

8. Find the radius of circle O if $k = 1$ and make an illustrative drawing.

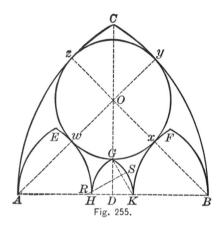

Fig. 255.

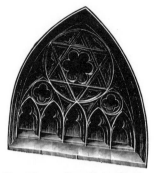

Fig. 255*a*. — From Union Congregational Church, Boston, Mass.

301. In Fig. 255 D is the middle point of AB. AH = BK = $\frac{2}{5}$ of AB. A is the center for the arcs EH and CB, B for the arcs KF and CA, H and K for the arcs EA and FB, respectively.

EXERCISES

1. Prove that arcs CA and AE are tangent; also FB and CB.

2. If circle O is tangent to the arcs CA and CB or to arcs EH and FK, prove that its center must lie on the line CD.

3. If circle O is tangent to the concentric arcs EH and CB, where must its center lie?

4. If circle O is tangent to the concentric arcs CA and FK, where must its center lie?

5. Show how to construct circle O.

Suggestion. — It is necessary to prove that $Ox = Oy = Oz = Ow$, and that circle O is tangent to the four arcs AC, CB, EH, and KF.

6. Show that the radius of circle O may be derived from the formula obtained in § 299, Ex. 5.

7. Prove that the points $A, w, O,$ and y, also $z, O, x,$ and B are collinear.

8. Construct the arc GK, passing through the points G and K with its center on line AB.

9. Show that the construction required for Ex. 8 is equivalent to constructing a circle passing through a given point and tangent to a given line at a given point.

10. Show how to construct the complete figure.

11. If $AB = 15$, find (1) the radius of circle O; (2) the lengths of OD and GD; (3) the area of curved triangle AEH.

12. If $AB = s$, find the lengths and areas mentioned in Ex. 11.

Ans. (1) $\dfrac{3\,s}{10}$; (2) $\dfrac{s}{5}\sqrt{6}$ and $\dfrac{s}{10}(2\sqrt{6}-3)$; (3) $\dfrac{s^2}{75}(4\pi - 3\sqrt{3})$.

13. If R is the center for the arc GK, show that $RD = \dfrac{\overline{GD}^2 - \overline{KD}^2}{2\,KD}$.

Suggestion. — $\triangle$ RSK and GDK are similar, therefore $\dfrac{RK}{GK} = \dfrac{SK}{DK}$ or $RK = \dfrac{GK \cdot SK}{DK}$. S is the mid-point of line GK. $RD = RK - DK$.

14. If $AB = s$, find DR. $\qquad\qquad$ *Ans.* $\dfrac{s}{5}(8 - 3\sqrt{6})$.

15. If $AH = KB = kAB = ks$, find (1) the radius of circle O; (2) the length of OD, DG, and DR; (3) the area of curved triangle AEH.

Ans. (1) $\dfrac{s}{2}(1 - k)$; (2) $\dfrac{s}{2}\sqrt{k^2 + 2\,k}$, $\dfrac{s}{2}(k - 1 + \sqrt{k^2 + 2\,k})$,

$\dfrac{s}{2}\left(\dfrac{2\,k - k^2 + (k - 1)(\sqrt{k^2 + 2\,k})}{1 - 2\,k}\right)$; (3) $\dfrac{s^2 k^2}{12}(\tfrac{1}{?}\pi - 3\sqrt{3})$.

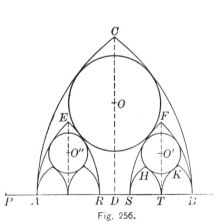

Fig. 256.

Fig. 256*a*. — Lincoln Cathedral.

302. In Fig. 256 $AQ = BP = \tfrac{6}{5}$ AB. P and Q are the centers for the arcs CB and AB, respectively. AB is divided into five equal parts. Arcs EA, ER, FS, and FB are drawn from S, P, Q, and R as centers and radius $\tfrac{3}{5}$ AB.

1. Find radius of circle O tangent to arcs AC, FS, CB, and ER.

$Ans.\ \dfrac{3\,s}{10}$.

Suggestion. — Notice that the arcs AC and FS, also CB and ER are concentric.

2. Show that the result found in Ex. 1 may be derived from the formula in § 299, Ex. 5.

3. Find the radius of circle O' tangent to the arcs FS and FB, and to the arcs HT and TK, if the arches SHT and TRB are equilateral.

$Ans.\ \dfrac{s}{8}$.

4. Show how the answer to Ex. 3 may be derived from the formula obtained in § 275, Ex. 23.

5. If the arches BKT, THS, etc., in Fig. 256 had been semi-circles, what would have been the radius of circle O', if $AB = s$?

$Ans.\ \dfrac{s}{7}$.

6. Derive the answer to Ex. 5 from the results obtained in § 278, Ex. 3.

Suggestion. — From § 278, Ex. 3, $r = \dfrac{SB}{2} \cdot \dfrac{4\,k - 1}{4\,k + 1}$; in Fig. 256 SB $= \dfrac{2\,s}{5}$; $k = \dfrac{3\,SB}{2}$. Make the necessary substitutions.

7. Draw a diagram to illustrate the case suggested in Ex. 5.

303. Type II.   Figures 257 and 258 are illustrations of Type II.   They may be referred to Type III of three-light windows (§ 294) as indicated in Ex. 2 below.

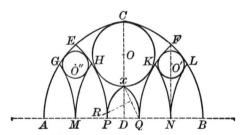

Fig. 257. — From Chester Cathedral Presbytery.   Bond, p. 524.

In Fig. 257 $AQ = BP = \frac{3}{5} AB$. P and Q are the centers for the arcs CB and AC, respectively. AB is divided into 5 equal parts. All arcs shown except xP and xQ are drawn with AQ as radius and centers on line AB or AB extended.

EXERCISES

1. Find radius of circles O and O' placed as shown in the figure, if $AB = s$. *Ans.* (1) $\frac{s}{5}$; (2) $\frac{s}{12}$.

2. Show that the radius of circle O can be obtained from the formula in § 294, Ex. 3, and the radius of circle O' from the formula in § 281, Ex. 3.

3. Construct the entire figure.

Suggestion. — Arcs xP and xQ are constructed to pass through points x and P, and x and Q, respectively, and to have centers on the line AB.

4. If $AB = s$, find the length of OD and Dx.

$$Ans. \ (1) \ \frac{s}{10}\sqrt{15}; \ (2) \ \frac{s}{10}(\sqrt{15} - 2).$$

5. If $AB = s$, find the distance RD, when R is the center for the arc xQ. *Ans.* $\frac{s}{10}(9 - 2\sqrt{15})$.

Suggestion. — As in § 301, Ex. 13, $RD = \dfrac{\overline{xD}^2 - \overline{DQ}^2}{2\,DQ}$.

6. If $AQ = kAB = ks$, find the radius of circle O'.

$$Ans. \ \frac{s}{100} \cdot \frac{10\,k - 1}{k}.$$

7. If $AQ = AB = s$, find the radii of circles O and O', and construct the entire figure. *Ans.* (1) $\frac{6\,s}{25}$; (2) $\frac{9\,s}{100}$.

Suggestion. — For the radius of circle O, the formula obtained in § 294, Ex. 3, may be used.

8. If $AQ = kAB = ks$, find the radii of circle O.

$$Ans. \ \frac{3\,s}{50} \cdot \left(\frac{5\,k - 1}{k}\right)$$

9. Construct a figure to illustrate the case $k = 1$.

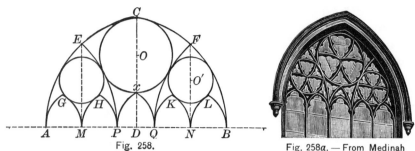

Fig. 258.

Fig. 258a. — From Medinah
Temple, Chicago.

304. **Figure 258 is like Fig. 257 except that arches AGM, MHP, QKN, NLB are equilateral.**

EXERCISES

1. Find the radius of circle O' if $AQ = BP = \frac{3}{5} AB = \frac{3}{5} s$.

Ans. $\dfrac{s}{8}$.

2. Construct the entire figure.

Suggestion. — The radius of circle O can be derived from § 294, Ex. 3.

3. If $AQ = kAB = ks$, find the radius of circle O'.

Ans. $\dfrac{s}{5} \cdot \dfrac{10\,k - 1}{10\,k + 2}$.

4. Show that the formula obtained in Ex. 3 may be derived from that in § 275, Ex. 23.

5. If $AQ = \frac{4}{5} AB = \frac{4}{5} s$, find the radius of circle O and O', and draw an illustrative figure. *Ans.* (1) $\dfrac{9\,s}{40}$; (2) $\dfrac{7\,s}{50}$.

6. If the arches AGM, MHP, etc., are semicircles, and if $AQ = \frac{3}{5} AB$, what is the radius of circle O'? *Ans.* $\dfrac{AB}{7}$.

7. Draw a figure to illustrate Ex. 6, using $AB = 7$ in.

8. If the arches AGM, MHP, etc., are semicircles, and if $AQ = kAB = ks$, find the radius of the circle O'. *Ans.* $\dfrac{s}{5} \cdot \dfrac{10\,k - 1}{10\,k + 1}$.

9. Show that the results obtained in Exs. 6 and 8 may be found from § 278, Ex. 3.

SEVEN-LIGHT WINDOWS

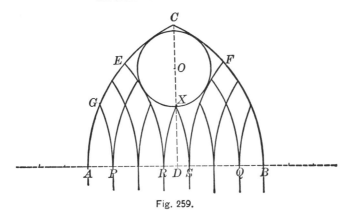

Fig. 259.

305. In Fig. 259 $AQ = BP = \frac{6}{7} AB$. P and Q are the centers for the arcs CB and AC. AB is divided into 7 equal parts. The arcs shown passing through the various points of division have radii equal to AQ and centers on AB or AB extended.

EXERCISES

1. Find the radius of circle O tangent to the arcs AC, CB, ER, and FS, if $AB = s$.   *Ans.* $\dfrac{3\,s}{14}$.

Fig. 259a. — From Merton College Chapel, Oxford.

2. Construct the entire figure.

3. Verify the result derived in Ex. 1 by substituting $k = \frac{6}{7}$, $m = \frac{3}{7}$, in the result obtained in § 294, Ex. 3.

4. If $AB = s$, find the length of XD and the altitude of the arch AGP.   *Ans.* (1) $\frac{s}{14}(2\sqrt{14} - 3)$; (2) $\frac{s}{14}\sqrt{23}$.

5. If $AQ = kAB = ks$, find the radius of circle O.

$$Ans.\ \frac{s}{49} \cdot \frac{14\,k - 3}{k}.$$

6. Verify the above result by substituting $m = \frac{3}{7}$ in the formula obtained in § 294, Ex. 3.

306. In Fig. 260 AQ = BP = $\frac{6}{7}$ AB. P and Q are the centers for arcs CB and CA respectively. AB is divided into 7 equal parts. Arcs ER and FS are drawn with radius AQ and center on line AB extended. Arches NLB, SKQ, etc., are equilateral on the lines NB, SQ, etc. respectively.

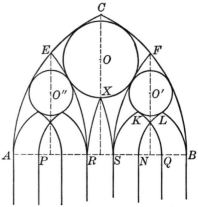

Fig. 260. — From Exeter Cathedral, Lady Chapel. See Bond p. 473.

EXERCISES

1. Find the radius of circle O' tangent to arcs FS, FB, KQ, and LN, if $AB = s$.   *Ans.* $\frac{s}{8}$.

2. Construct the entire figure, using $AB = 7$ in.

Suggestion. — The radius of circle O can be found from formula given in § 294, Ex. 3. Arcs XS and XR are constructed to pass through the points X and R, and S and R, respectively, with centers on line AB.

3. If $AQ = kAB = ks$, find the radius of circle O'. *Ans.* $\frac{s}{14} \cdot \frac{21\,k - 4}{7\,k + 2}$.

4. Show that the answer to Ex. 3 may be derived from the formula obtained in § 289, Ex. 4.

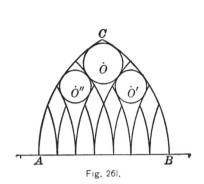

Fig. 261.

Fig. 261a. — From Tintern Abbey.

307. In Fig. 261 the arch ACB is equilateral. AB is divided into 7 equal parts, and the arcs drawn as shown.

EXERCISES

1. Find the radius of circle O and of circle O', if $AB = s$.

$$Ans. \quad \frac{15\,s}{98}, \quad \frac{6\,s}{49}.$$

2. Construct the entire figure.

3. Show that the radius of circle O may be found from the formula given in § 294, Ex. 3.

4. Show that the radius of circle O' may be found from the formula given in § 294, Ex. 3.

5. Show that the radius of circle O' may be found from the formula given in § 281, Ex. 3.

6. If the radius for the arcs CA and CB is kAB, find the radii of circles O and O'.

PART 4. VENETIAN TRACERY

308. Venetian tracery is peculiar in that it is intended for indefinite horizontal extension, and not designed to fit a definite closed space.[1] For a modern example of the same see the Montauk Club, Brooklyn, N. Y.[2]

---

[1] Brochure series, January, 1895, p. 7.   [2] *Arch. Record*, Vol. II, p. 136.

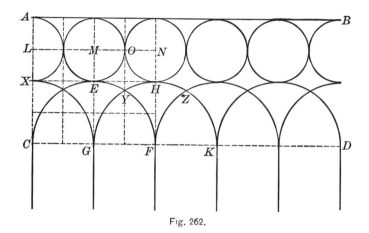

Fig. 262.

309. In Fig. 262 circles L, M, N, etc., are tangent to the line AB, to each other, and to the semicircles drawn with C, G, F, K, etc., as centers.

EXERCISES

1. Prove that $AX = XC$ and that $AL = LX$, if AB and CD are parallel and E, H, etc., are midpoints of the arcs CEF, GHK, etc.

2. Given the parallel lines AB and CD. Construct the figure.

3. If $AC = 6$, find the area of the following curved figures:

   (*a*) $\triangle GYF$.   *Ans.* 5.52+.

   (*b*) Triangle $YFHZ$.

                 Ans. 3.08+.

   (*c*) Quadrilateral $EOHY$.

                 Ans. 1.35+.

4. If $AC = h$, find the areas mentioned in Ex. 3.

Ans.   (*a*) $\dfrac{h^2}{48}(4\pi - 3\sqrt{3})$;

      (*b*) $\dfrac{h^2}{24}(3\sqrt{3} - \pi)$;

      (*c*) $\dfrac{h^2}{96}(36 - 7\pi - 6\sqrt{3})$.

Fig. 262*a*. — From Franchetti Palace, Venice, Italy.

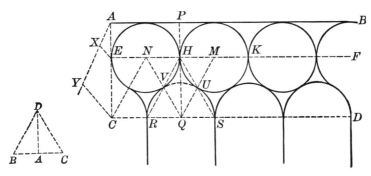

Fig. 263.

310. In Fig. 263 AB is parallel to CD, the equal circles M and N are tangent to each other at H and to the line AB, and the semicircle with Q as its center is tangent to circle N at V and to circle M at U.

EXERCISES

1. If line QH is extended to meet AB at P, find the ratio $\dfrac{PH}{HQ}$.

Suggestion. — Prove $\triangle NMQ$ equilateral with the points of contact H, U, and V on the sides of the triangle, and $HN = HM = HP$. Therefore $\dfrac{HP}{HQ} = \dfrac{1}{\sqrt{3}}$.

2. Given the two parallels AB and CD. Construct the series of circles and semicircles as indicated.

Suggestion. — Divide PQ at H in the ratio of the altitude of an equilateral triangle to one half its side. Through H draw EF parallel to AB. Construct the circles and semicircles as indicated. Prove the semicircles tangent to the circles.

3. If $PQ = 6$, find the area of the curved figure $RVHUS$.

Ans. $18(2\sqrt{3} - 3)$.

4. If $PQ = a$, find the same area. Ans. $\dfrac{a^2}{2}(2\sqrt{3} - 3)$.

5. If $PQ = 6$, find the length of the curve RVH.

6. If $PQ = a$, find the length of the same curve.

7. If $PQ = 6$, find the area of the cusped triangle VUH.

8. If $PQ = a$, find the same area.

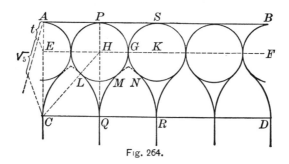

Fig. 264.

Fig. 264a. — From the Doge's Palace, Venice.

311. In Fig. 264 lines AB and CD are parallel and the circles E, H, and K are tangent to each other and to the equilateral arches, CLQ, QMR, etc.

1. Prove that (1) the radius of circle H is equal to $\frac{1}{2} CQ$; (2) PQ is perpendicular to AB and CD; (3) Q, H, and P are collinear.

2. Find the ratio $\dfrac{PH}{HQ}$.

Suggestion. — $PH = HL.$ $HC = 3\, HL.$ $CQ = 2\, HL.$ Hence $\dfrac{PH}{HQ} = \dfrac{1}{\sqrt{5}}.$

3. Given the parallel lines AB and CD. Construct the series of tangent circles and equilateral arches as indicated.

CHAPTER VI

TRUSSES AND ARCHES

PART 1. RAFTERED ROOFS AND TRUSSES

312. Definition. — Raftered roofs are classified according to construction into single-framed or untrussed roofs, and double-framed or trussed roofs. In double-framed roofs the ordinary rafters are carried by sets of supporting timbers or by steel frames called trusses. These are placed eight or nine feet apart, and carry horizontal bars called purlins which support the common rafters.

313. Development of the Truss. — The isosceles triangle is the simplest form of truss and contains all elements really essential. In spanning large spaces the tie beam (AB, Fig. 265) may be so long and heavy as to sag of its own weight. The king-post (CD, Fig. 265) is then added as support to the tie beam. Rafters over 12 feet long are also liable to sag. Hence the addition of struts or braces (DE and DF, Fig. 265). As the space to be spanned becomes still larger, trusses become more and more complicated. Figure 266 shows the queen posts EF and HK and 4 braces. Figure 268 shows the princess posts, KL, PQ, etc., additional braces, and an attic space between the queen posts EG and FH. For a complete study of the problem of the truss, graphic statics is necessary. All parts must be in equilibrium under the weights to be supported.

The forms shown, with the exception of Figs. 270 and 271, may be made wholly or partly of wood. Those shown in Figs. 270 and 271 are steel types. The earliest steel types were like the wooden ones; later, more economical designs were introduced, many of which were originally suggested by the stress and strain diagrams.

314. In Fig. 265 ABC is an isosceles triangle. CD is the median to the base AB. DE and DF are drawn from D to the middle points of the lines AC and BC, respectively.

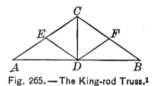

Fig. 265. — The King-rod Truss.[1]

EXERCISES

1. Prove that (1) CD is perpendicular to AB, (2) DE is $\frac{1}{2} AC$, and (3) $\triangle ADE$ is $\frac{1}{4}$ the $\triangle ABC$.

2. If $CD = \frac{1}{4} AB$ and AE is 12 ft., find the length of AB.

Ans. 42.93+.

3. If the pitch of the roof is 7 in. to the foot and AE is 12 ft., find AB.

315. In Fig. 266 ABC is an isosceles triangle. CD is drawn from C to middle point of AB. AG = GE = EC and CH = HL = LB. DF = $\frac{1}{3}$ AD = DK. The points are joined as indicated.

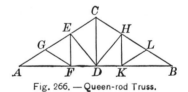

Fig. 266. — Queen-rod Truss.

EXERCISES

1. Prove that

(*a*) EF is perpendicular to AB; (*b*) $EF = HK$; (*c*) $GF = GA$; (*d*) $\triangle FDE$ has same area as $\triangle FEG$; (*e*) the area of $\triangle FDE$ is $\frac{2}{3}$ the area of $\triangle DCE$.

---

[1] The names of trusses follow Kidder (2).

Suggestion for (*e*). — Consider *EF* and *DC* as the bases; then the triangles have equal altitudes and are to each other as their bases. Prove that $EF = \frac{2}{3}DC$.

2. If $GE = 12$ and $CD = \frac{1}{3}AB$, find (1) *AB*, (2) *CD*, (3) *EF*, (4) *ED*. *Ans.* (1) 59.90; (2) 19.96; (3) 13.31; (4) 16.64.

Suggestion for (4). — By use of radicals $FD = \frac{36}{13}\sqrt{13}$ and $EF = \frac{48}{13}\sqrt{13}$. Squaring and adding $\overline{ED}^2 = \frac{3600}{13}$.

316. In Fig. 267 ABC is an isosceles triangle. AG = GF = FC and CL = LK = KB. FD = DL. FE and LH are perpendicular to AB, and the points are joined as indicated.

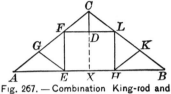

Fig. 267. — Combination King-rod and Queen-rod Truss.

EXERCISES

1. Prove that

(1) *CD* is perpendicular to *FL* and, if extended, to *AB*.

(2) *CD* if extended would bisect *AB*.

(3) $GE = AG$.

(4) *FL* is parallel to *AB* and equal to $\frac{1}{3}AB$.

(5) $FE = LH$.

(6) The areas of △ *AEG* and *GEF* are equal, and equal to $\frac{1}{4}$ the area of *EHLF*.

2. Find the positions of points *F* and *L* so that *EHLF* is a square.

3. What must be the relation between *CX* and *AB* if *EHLF* is a square and *CL* is $\frac{1}{3}CB$? *Ans.* $CX = \frac{1}{2}AB$.

Suggestion. — If $CL = \frac{1}{3}CB$, *CD* is $\frac{1}{3}CX$. Therefore $CD = DL$, and hence $CX = XB$.

4. What must be the relation between *CX* and *AB* if *EHLF* is a square and $CL = \frac{1}{4}CB$? *Ans.* $CX = \frac{1}{3}AB$.

5. If $LH = \frac{2}{3} FL$ and $CL = \frac{1}{3} AB$, find the relation between CX and AB. *Ans.* $CX = \frac{1}{3} AB$.

6. If $CX = \frac{1}{4} AB$, find the relation between FL and LH, so that $CL = \frac{1}{3} CB$. *Ans.* $FL = 2 LH$.

7. Given the $\triangle ABC$; find points F and L so that $FL = 3 LH$.

317. In Fig. 268 ABC is an isosceles triangle. $AK = KE = EP = PC$ and $CR = RF = FM = MB$. $ED = DF$. KL, EG, FH, MN are perpendicular to AB. PQ and RS are perpendicular to EF.

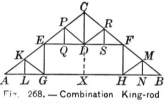

Fig. 268. — Combination King-rod and Queen-rod Truss.

EXERCISES

1. Prove that

(1) CD is perpendicular to EF.

(2) GK and DP are equal and equal to $\frac{1}{4} AC$.

(3) EF is parallel to AB and equal to $\frac{1}{2} AB$.

(4) KG and DP are parallel.

(5) The area of $\triangle EFC$ is equal to $\frac{1}{2}$ that of $EGHF$.

(6) The area of $\triangle ALK$ is $\frac{1}{32}$ the area of $\triangle ABC$.

2. If $CX = XB$, prove that $HF = FD$.

3. If $CF = FB$ and $\dfrac{CX}{XB} = \dfrac{5}{12}$, find the relation between FH and FE. *Ans.* $\frac{5}{24}$.

4. If $CF = FB$ and $FH = \frac{1}{4} EF$, find the ratio $\dfrac{CX}{XB}$. *Ans.* $\frac{1}{2}$.

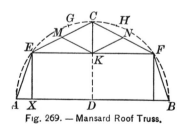

Fig. 269. — Mansard Roof Truss.

318. 1. In Fig. 269, if EF is parallel to AB and $CE = CF$, prove that

(*a*) $\angle A = \angle B$.

(*b*) The line bisecting $\angle C$ is the perpendicular bisector of AB and EF.

(*c*) The lines BE and AF are equal.

2. The outline for the truss may be determined as follows : Draw semicircle ABC and divide it into five equal parts at points F, H, G, and E. Let C be the center of arc GH. Join points as indicated.[1] In this case prove that

(*a*) EF is parallel to AB.

(*b*) CK is perpendicular to EF and bisects $\angle C$, if K is center of EF.

(*c*) CK extended bisects AB at right angles.

3. If the outline of the truss is constructed as in Ex. 2, find the number of degrees in $\angle EAB$, $\angle AEC$, $\angle CEK$, $\angle ECF$.
$Ans.$ (1) $72°$; (2) $135°$; (3) $27°$; (4) $126°$.

4. A second method for constructing the outline of the truss for any given height and span is as follows: At the ends of the span construct $\angle s$ A and B each $60°$ ($60°$ to $70°$ in practice). Draw the line EF parallel to AB at the desired height above AB. Construct $\angle s$ CEF and CFE at points E and F on line EF each $30°$.[1] Construct the figure according to the above method to scale, $\frac{1}{10}$ inch to foot. $AB = 30$ ft., $EX = 10$ ft.

5. Can the figure constructed as in Ex. 4 be inscribed in a semicircle?

6. If $AB = 30$, $EX = 10$, $\angle A = 60°$, and $\angle CEK = 30°$, find the length of AE, EF, CK, and EC.
$Ans.$ (1) $\frac{20}{3}\sqrt{3} = 11.54$; (2) 18.45; (3) 5.32; (4) 10.65.

7. If $EM = MC = 8$, find the length of AB, if $\angle A = 60°$, $\angle CEK = 30°$, and $EX = 10$. $Ans.$ 39.26.

[1] Cassell's, p. 134.

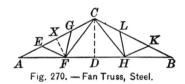

Fig. 270. — Fan Truss, Steel.

319. 1. In Fig. 270 ABC is an isosceles triangle. Show how to construct the figure so that $AF = FC = CH = HB$ and $AE = EF = FG = GC$.

2. If the ratio $\dfrac{CD}{AD} = \dfrac{5}{12}$ and $AB = 40$, find the length of CF and EF.   *Ans.* (1) 11.73; (2) 6.3.

Suggestion for CF. — Show that $AD = 20$ and $CD = 8\frac{1}{3}$. Let $AF = x$. Solve for x in the equation $x^2 - (20 - x)^2 = (\frac{25}{3})^2$.

Suggestion for EF. — Show that $XF = \frac{352}{72}$. Solve for x in the equation, $x^2 - (\frac{65}{6} - x)^2 = (\frac{325}{72})^2$.

3. Prove that ⧌ ABC, AFC, and AFE are similar and verify the results obtained in Ex. 2 by means of the similar ⧌.

320. In Fig. 271 ABC is an isosceles triangle. CH and CQ are so constructed that AH = HC = CQ = QB. F and K are the middle points of AH and HC, respectively, and AE = EG = GL = LC.

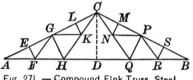

Fig. 271. — Compound Fink Truss, Steel.

1. Prove that

(1) FE, HG, and KL are perpendicular to AC.

(2) GK is parallel to AD.

(3) $FHKG$ is a rhombus.

2. If P and N are the middle points of CB and CQ, respectively, prove that the points G, K, N, and P are collinear.

3. If $\dfrac{CD}{AD} = \dfrac{1}{2}$ and $AB = 50$, find the length of CH, HG, and EF.

Ans. (1) $15\frac{5}{8}$; (2) $\frac{25}{8}\sqrt{5}$; (3) $\frac{25}{16}\sqrt{5}$.

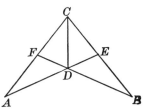

Fig. 272. — Scissors Truss

EXERCISES

321. 1. In Fig. 272, if $CA = CB$ and $CF = CE$, prove that $AE = FB$ and that $DA = DB$.

2. Join C and D. Prove that CD bisects $\angle C$ and is the perpendicular bisector of AB.

3. If $CA = CB$, and AE and BF are perpendicular to CB and CA, respectively, prove that $AE = BF$ and that $AD = DB$.

4. If $\angle C = 60°$, if AE and BF are perpendicular to CB and AC, respectively, and if $CA = 16$, find AE, AD, and CD.

Ans. (1) $8\sqrt{3}$; (2) $\tfrac{16}{3}\sqrt{3}$; (3) $\tfrac{16}{3}\sqrt{3}$.

PART 2. ARCHES

322. History. — The arch first appeared as an architectural feature in the valleys of the Tigris and Euphrates rivers. Here the alluvial soil, which contained no stone and bore no trees, forced the use of brick for building material and made the use of stone and wooden lintels impossible. The discovery of the arch naturally followed. Semicircular and pointed forms are said to date as far back as the eighth or ninth century B.C.[1] No form of the arch was in general use in Europe until Roman times. In Greece stone available for building material was abundant, and consequently Greek architecture is based on the lintel.

The general introduction of the arch, the vault, and the

[1] Fergusson, Vol. I, p. 215 ; Fletcher, p. 43.

dome as architectural features is due to the Romans. These forms had two immense advantages over the lintel : they were adapted to all kinds of building material ; they did away at once with the crowd of pillars that filled rooms and halls when short stone lintels were used. Because of these advantages carried throughout the Roman dominions, these new architectural features were soon established as essential to all building. The outline of all Roman forms was semicircular.

Later other forms of the arch came into use as necessity demanded. At first these were formed of the arcs of circles as shown in the following sections. In the best modern construction, elliptical, hyperbolic, parabolic, and cycloidal arches have replaced the earlier and cruder forms. As the drawings necessary to determine the shapes of the stones when the arches are true mathematical curves are much more difficult than is the case when they are combinations of arcs of circles, the latter are still used in all but the very best work.

POINTED OR GOTHIC ARCHES

323. History. — The general introduction of the pointed arch was due to the medieval architects of western Europe. While this arch may possibly have been borrowed from the Saracens, who up to this time had used it for ornamental purposes, yet the problems in vaulting encountered by the Gothic builders made its independent discovery and use almost certain.[1] In the earliest buildings in England in which it is found it is used for structural purposes only. Among these buildings is Gloucester Cathedral.[2]

[1] Fletcher, pp. 272 and 283 ; Fergusson, Vol. II, p. 45.
[2] Bond, p. 266.

324. Gables and canopies over doors and windows abound in Gothic buildings. Pierced and filled with tracery they were quite characteristic of fourteenth-century work. Later they became very elaborate.[1]

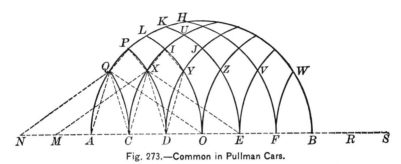

Fig. 273.—Common in Pullman Cars.

325. In Fig. 273 AB is the diameter of the semicircle AHB. AB is divided into six equal parts. NM = MA = AC = BR, etc. The arcs DP, LO, KE, etc. are drawn with M, A, C, etc. as centers and ½ AB as radius.

EXERCISES

1. Prove that AQC, CXD, DYO, etc. are congruent isosceles triangles.

2. Prove that the curved triangles AQC, CXD, DYO, etc. are congruent.

3. Prove that the linear △ APD, CIO, DJE, etc. are congruent and isosceles; also that the curved △ APD, CIO, DJE are congruent.

4. Prove by superposition that the curved quadrilaterals $PQCX$, $IXDY$, etc. are congruent.

5. If this design is to be executed in leaded glass, how many different shaped patterns must be cut? Prove.

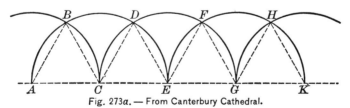

Fig. 273a. — From Canterbury Cathedral.

[1] Sturgis (1) pp. 273, 330.

6. Show that the equilateral Gothic arch may be derived from a series of intersecting semicircles.   (See Fig. 273*a*.)

Suggestion. — Prove ▵ *ABC*, *CDE*, *EFG* ⋯ etc. equilateral.

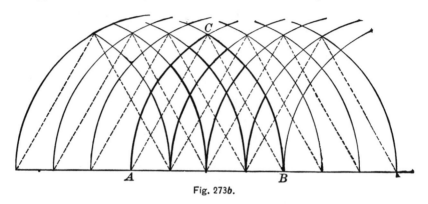

Fig. 273*b*.

7. Show that designs like those in Figs. 273 and 273*b* may be considered as based on a series of intersecting equilateral arches, or on a series of intersecting semicircles.

326. **In Fig. 274 AB and CD are of given lengths. CD is the perpendicular bisector of AB.**

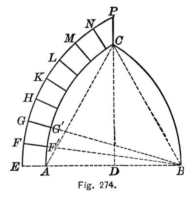

Fig. 274.

EXERCISES

1. If $\overset{\frown}{AC}$ and $\overset{\frown}{CB}$ are so constructed that their centers are on line *AB* and that they pass through the points *A* and *C*, and *B* and *C*, respectively, prove that they have equal radii.

2. If $\overset{\frown}{AC}$ and $\overset{\frown}{BC}$ are constructed as in Ex. 1 and if *A* is the center for $\overset{\frown}{BC}$, prove that △ *ABC* is equilateral.

3. Construct the drawing for eight stones on the side AC so that the extrados EF, FG, GH, etc. shall be equal.

Suggestion. — $\overparen{EP}$ and $\overparen{AC}$ must be concentric and the sutures FF', GG', etc., must be on radii. Therefore join P and B, and divide the angle EBP into eight equal parts.

4. How far are problems like Ex. 3 possible by elementary geometry?

Ans. When the number of stones required is 2^n.

Remark. — The true Gothic arch has no keystone.

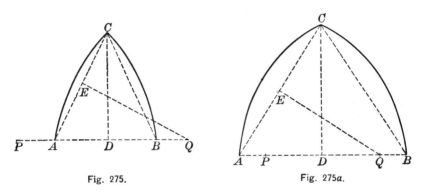

Fig. 275.   Fig. 275a.

327. In Figs. 275 and 275a, CD is the perpendicular bisector of AB. P and Q are the centers of the arcs CB and AC, respectively.

EXERCISES

1. If $AQ < AB$, prove that $AC < AB$. If $AQ > AB$, prove that $AC > AB$.

Suggestion. — Let E be the center of line AC. $\triangle AEQ$ and CAD are similar. Therefore $\dfrac{AQ}{AC} = \dfrac{AE}{AD}$.

From this it follows that $\dfrac{AQ}{AC} = \dfrac{AC}{AB}$ or $AQ \cdot AB = \overline{AC}^2$. Hence the relations given above.

2. Draw a figure for a pointed arch in which the altitude is $\frac{1}{2}$ the span. Draw a figure in which the altitude is less than $\frac{1}{2}$ the span. What must be the relation between the altitude and span in a true Gothic or pointed arch?

3. If $AB = s$ and $CD = h$, find the length of AQ.

$$Ans. \ \frac{s^2 + 4\,h^2}{4\,s}.$$

Suggestion. — Use the similar ⧌ AEQ and ACD.

4. If $AB = 8$ and $CD = 5$, find the length of AQ. $Ans. \ 5\frac{1}{8}.$

5. If $AQ = 6\frac{1}{4}$, and $CD = 6$, find the length of AB.

$$Ans. \ 9 \ or \ 16.$$

Suggestion. — Solve the equation $\dfrac{25}{4} = \dfrac{s^2 + 144}{4\,s}$ for s.

6. Construct a figure to scale to illustrate each answer obtained in Ex. 5.

7. If $AB = s$ and $AQ = r$, find CD. $Ans. \ \frac{1}{2}\sqrt{4\,rs - s^2}.$

8. If $AQ = r$ and $CD = h$, find AB. $Ans. \ 2\,r \pm 2\sqrt{r^2 - h^2}.$

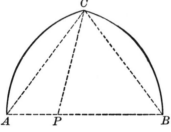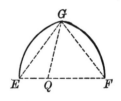

Fig. 276.

328. In Fig. 276 P and Q are respectively centers for the arcs BC and GF, in the arches ACB and EGF.

EXERCISES

1. If $\dfrac{BP}{AB} = \dfrac{QF}{EF}$, prove that ⧌ ABC and EFG are similar.

Suggestion. — Prove ⧌ ABC and PBC similar. Hence $AB \cdot PB = \overline{BC}^2$. Let $\dfrac{BP}{AB} = m$. Therefore $BP = mAB$ and $m\overline{AB}^2 = \overline{BC}^2$. Similarly $m\overline{EF}^2 = \overline{GF}^2$. Hence $\dfrac{BA}{EF} = \dfrac{CB}{FG}$. Therefore, since the triangles are isosceles, they are similar.

2. Draw a figure in which PB is greater than AB and prove Ex. 1 for this figure.

GABLES AND CANOPIES

329. In Fig. 277 and 277*a*, ACB is a Gothic arch formed of the two equal intersecting circles whose centers are H and K. DC is the common chord. MPL is a gable over the arch with its sides PL and PM drawn from a point in DC extended and tangent to the sides of the arch at L and M, respectively.

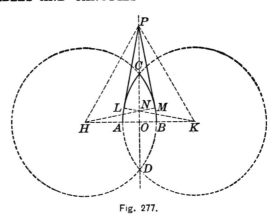

Fig. 277.

EXERCISES

1. Show how to construct the gable over a given Gothic arch, if point *P* is given.

2. Prove that $PL = PM$.

3. Prove that $\angle OPL = \angle OPM$ when (*a*) AB, the span of the arch, is less than HK, the line of centers, (*b*) $AB > HK$, (*c*) $AB = HK$.

Suggestion for (*a*).— See Fig. 277. Prove $\triangle HPM = \triangle PLK$ and $\triangle HNP = \triangle KNP$. Construct a figure to illustrate case (*b*).

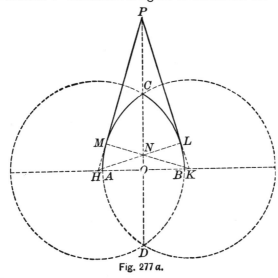

Fig. 277 *a.*

4. Show how to construct the gable over a given arch, if the length of one side of the gable is given.

Suggestion. — With the radius of the circle and the given length as legs construct a right triangle. The hypothenuse will be the length of *KP*. Locate point *P*.

5. Show how to construct the gable over a given arch, if its angle *LPM* is given.

Suggestion. — At any point in the line *PN* construct an angle equal to ½ the given angle. Then draw a tangent to the circle which shall be parallel to one side of this angle. Is this problem always possible? Is there any ambiguity in the construction?

6. Construct a figure in which *PH* and *PK* fall upon *PM* and *PL*, given the radius of the circles that form the arch and the span of the arch.

Suggestion. — See Fig. 277*a*. Is the problem always possible?

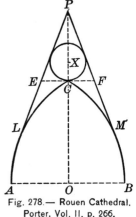

330. In Fig. 278 ACB is an equilateral Gothic arch. MPL is a gable above the arch. Circle O is tangent to PL and to PM and passes through point C.

Fig. 278.— Rouen Cathedral.
Porter, Vol. II, p. 266.

EXERCISES

1. Show how to construct circle *X*.

2. Construct a figure similar to Fig. 278, making the arch equilateral. Let $AB = 4$ in., $CP = 3$ in.

3. Construct a figure similar to Fig. 278, making $AB = 5$ in., $CO = 4\frac{1}{2}$ in., and $PL = 5$ in.

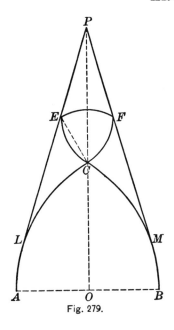

Fig. 279.

Fig. 279a. — From Church of Our
Lady of Lourdes, New York City.

331.— In Fig. 279 ACB is an equilateral arch with the gable MPL. CFE is an equilateral circular triangle struck from the centers C, F, and E.

EXERCISE

Show how to construct the entire figure.

THE POINTED ARCH IN THE SQUARE FRAME

332. In Fig. 280 $\widehat{AC}$ and $\widehat{CB}$ are tangent to AE and FB, respectively. C is the center of line EF. Circles O and O' are tangent to the sides of the rectangle and to the arcs as indicated.

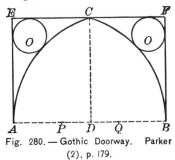

Fig. 280. — Gothic Doorway. Parker
(2), p. 179.

EXERCISES

1. Show how to find P and Q, the centers for the arcs BC and AC, respectively.

Suggestion. — See § 266, Ex. 3.

2. Prove that $AQ = PB$.

3. Construct circles O and O'.

Suggestion. — This involves the problem, " To construct a circle tangent to the sides of an angle and to a given circle." As the arc AC would cut the line EC if extended, three circles are possible in this particular application of the problem, and considerable care must be exercised in making that construction which gives the circle desired. See § 253.

4. If $AB = 8$ and $PB = 6$, find the length of CD.

Suggestion. — Draw the lines CB, CA, and CP. Use the similar ⧍ ACB and CPB.

OGEE ARCHES

333. Occurrence. — The ogee arch is much used on Turkish buildings, and is common as an ornamental feature in the late Gothic. It is very generally used in canopies and gables[1] in the masses of ornament that cover the inside and the outside of the later Gothic cathedrals. Half of it is sometimes used for veranda rafters, and for roofs of bay and oriel windows.

THREE–CENTERED OGEE ARCHES

334. In Fig. 281 the semicircle AGHB is constructed on AB as diameter, and CD is perpendicular to AB at its middle point.

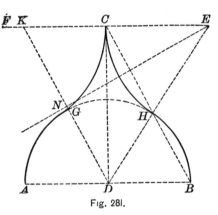

Fig. 281.

[1] Bond, p. 268.

1. Construct arcs CH and CG tangent to the line CD at point C, and to the semicircle.

Suggestion. — Make $KC = AD$. Draw NE, the perpendicular bisector of KD, meeting CK extended at E. E is center for the arc CH. What general problem in construction of circles is involved here?

2. If the arcs CH, drawn with E as a center, and HB drawn with D as a center, are tangent at H, prove that the points D, H, and E are collinear.

3. Prove that C, H, and B are collinear.

Suggestion. — Join C and H and B and C, and prove that each is parallel to DK.

4. If $AB = 8$ and CD is 8, find the length of CE.   *Ans.* 6.

Suggestion. — $KD = 4\sqrt{5}$. Compare the sides of the similar triangles KCD and KNE. Find KE and hence CE.

5. If $AB = s$ and $CD = h$, find the length of CE.

$$Ans. \ \frac{4\,h^2 - s^2}{4\,s}.$$

6. For what value of CD will $CE = DB$?   *Ans.* $CD = \dfrac{AB\sqrt{3}}{2}$.

Suggestion. — Let $AB = s$, then $CE = \dfrac{s}{2}$. Solve the equation for h,
$$\frac{s}{2} = \frac{4\,h^2 - s^2}{4\,s}.$$

7. Construct a figure to illustrate the case given in Ex. 6.

Suggestion. — Construct ACB an equilateral triangle.

8. If $CE = DB$, prove that $CH = HB$.

Suggestion. — See Ex. 3. Compare the similar $\triangle$ CEH and DHB.

9. If $CE = DB$ and $AB = s$, find the area of the curved figure $AGCHB$.   *Ans.* $\dfrac{s^2}{4}\sqrt{3}$.

10. If the span AB is given and if CD is of indefinite length, construct $\overset{\frown}{CH}$ so that it shall be tangent to CD and to the semicircle $AGHB$ and have a given radius.

Suggestion. — Point E may be found by the intersection of two loci, namely: (1) the locus of centers of circles of given radius and tangent to a given circle; (2) the locus of centers of circles of given radius and tangent to a given line.

11. If $AB = s$ and $CE = r$, find the length of CD.

$$Ans.\ \tfrac{1}{2}\sqrt{s(4r + s)}.$$

Suggestion. — Solve for h the equation, $r = \dfrac{4h^2 - s^2}{4s}$.

FOUR-CENTERED OGEE ARCHES

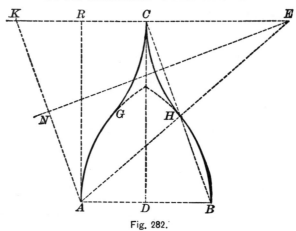

Fig. 282.

335. In Fig. 282 AB and CD are of given lengths. A is the center for the arc HB, B is the center for the arc AG. CD is the perpendicular bisector of the span AB.

EXERCISES

1. Construct the arcs CH and CG tangent to the arcs HB and AG, respectively, and to the line CD at C.

2. Draw lines AE and CB. Prove that they intersect at the point of tangency H.

3. If $AB = s$ and $CD = h$, find the length of CE.

$$Ans.\ \dfrac{4h^2 - 3s^2}{4s}.$$

Suggestion. — Find the value of KA and compare the sides of the similar ⧄ KRA and KEN.

4. For what value of CD will $CE = AB$? *Ans.* $CD = \dfrac{AB\sqrt{7}}{2}$.

Suggestion. — Let $CE = AB = s$. Solve for h in the equation $\dfrac{4\,h^2 - 3\,s^2}{4\,s} = s.$

5. If $CE = AB$, prove that H is the mid-point of the line CB.

6. What must be the length of CE if $CH = \frac{1}{2}\,HB$?
$$Ans.\ CE = \tfrac{1}{2}\,AB.$$

Suggestion. — Compare the sides of the similar ⧄ CEH and HAB.

7. Find the value of CD if $CE = \frac{1}{2}\,AB$.　　*Ans.* $CD = \dfrac{AB\sqrt{5}}{2}$.

8. What must be the length of CD if $CE = \frac{1}{3}\,AB$?
$$Ans.\ \dfrac{AB}{6}\sqrt{39}.$$

9. If the span AB is given and if CD is of indefinite length, construct $\overset{\frown}{CH}$ so that it shall be tangent to the line CD and to $\overset{\frown}{HB}$ and have a given radius.

Suggestion. — Find point E by intersecting loci as in § 334, Ex. 10.

10. By the construction suggested in Ex. 9, draw a figure in which $CE = AB$.

336. In Fig. 283 AB and CD are of given lengths. P and Q are fixed points on the line AB such that A Q = PB. P and Q are the centers for the arcs HB and GA, respectively. CD is the perpendicular bisector of AB.

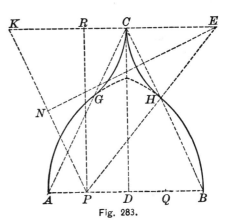

Fig. 283.

1. Construct the arcs CH and GC tangent to the line CD at C, and to the arcs HB and GA.

2. Draw lines PE and CB. Prove that they intersect at the point of tangency H.

3. If $AB = s$, $PB = ks$, and $CD = h$, find the length of CE.

$$Ans. \quad \frac{4\,h^2 + s^2(1 - 4k)}{4\,s}.$$

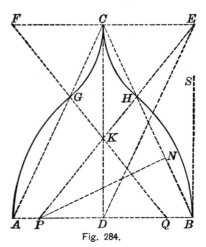

337. In Fig. 284 AB and CD are given lengths. H and G are fixed points on the sides of $\triangle$ ABC. CD is a perpendicular bisector of AB. BS is perpendicular to AB at B.

Fig. 284.

1. Construct the arcs CH and HB so that they shall be tangent to CD at C and to BS at B, respectively, and so that they shall pass through a given point H in the line CB.

2. If E and P are the centers for the arcs CH and HB, respectively, prove that the points E, H, and P are collinear and that the arcs CH and HB are tangent at H.

Suggestion. — To prove that the points E, H, and P are collinear, prove that $\angle\, CHE = \angle\, BHP$.

3. If $CG = CH$, construct the entire four-centered arch $AGCHB$, given the span AB, the altitude CD, and the points of inflection G and H on the sides of $\triangle ABC$ such that $CH = CG$.[1]

[1] Hanstein, Pl. 20, gives a construction which is a special case of the above.

4. If P and Q are centers for the arcs HB and GA, respectively, and E and F for CH and CG, respectively, prove that

(1) $EC = CF$.

(2) $PD = DQ$.

(3) EP, FQ, and CD are concurrent.

BASKET-HANDLE ARCHES

338. History. — Basket-handle arches, which were especially common in France,[1] seem to have arisen in late Gothic times, as arches lower than semicircles came to be used. They had one distinct advantage over arches formed of an arc of a single circle; in these latter forms, weight upon the top of the arch results in considerable outward pressure at the point where the arch springs from the supporting pillars, while in basket-handle arches a large component of the pressure at this point is downward.

339. Of the constructions given in this section, § 341 and § 342 contain those that are most generally applicable, as in them the center for one of the circles is arbitrary when the altitude and span are given. It may be readily seen by experiment that the relative position of the centers determines the slope of the arch for any given span and rise.

340. In Fig. 285 PQ, the line of centers of the two equal non-intersecting circles P and Q, is extended to meet the circles at A and B, respectively. CD is the perpendicular bisector of AB.

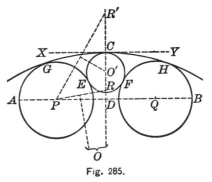

Fig. 285.

[1] Bond, p. 267.

EXERCISES

1. Prove that a circle which is tangent to circle P and has its center on line CD is tangent also to circle Q.

2. If XY is perpendicular to CD at C, construct a circle tangent to XY at C and to circle P. Show that two circles are possible.

3. Construct a figure for each of the following cases and note the relative position of circle O to circles P and Q.

(a) $CD > AD$, (b) $CD = AD$, (c) $CD = AP$, (d) $CD < AP$, (e) $AP < CD < AD$.

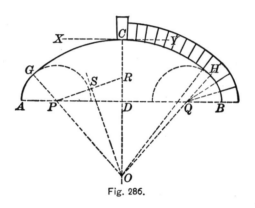

Fig. 286.

EXERCISES

341. **1.** Construct a basket-handle arch, given the span AB, the altitude CD, and the points P and Q to be used as centers for the arcs AG and HB.

Suggestion. — See § 340, Ex. 3 (e), the application of which is shown by the dotted lines in Fig. 286. CR should $= AP$.

2. Prove that $\triangle\ PDR$ and SRO are similar, and hence prove that $\angle RPD = \frac{1}{2} \angle POR$.

3. Prove that $\angle APG + 2 \angle RPD = 1$ rt. $\angle$. Hence invent another construction for determining point O.

4. If $AB = 14$, $CD = 4$, and $AP = 3$, find the length of PR, RO, and DO. *Ans.* (1) $\sqrt{17}$; (2) $\frac{17}{2}$; (3) $\frac{15}{2}$.

Suggestion. — Find RP, and use similar triangles PRD and SRO.

5. If $AB=s$, $CD=h$, and $AP=r$, find the length of PR, RO and CO.

$$Ans. \ (1) \ \sqrt{\left(\frac{s}{2}-r\right)^2+(h-r)^2}; \quad (2) \ \frac{\left(\frac{s}{2}-r\right)^2+(h-r)^2}{2(h-r)};$$

$$(3) \ \frac{s^2-4\,rs+4\,h^2}{8(h-r)}.$$

Suggestion for 2. — From the similar $\triangle\ PRD$ and SRO, $RO=\dfrac{2\,\overline{RS}^2}{RD}$.

6. If $AP=\dfrac{AB}{6}$ and $CD=\dfrac{AB}{3}$, find the length of CO.

$$Ans. \ \frac{7\,s}{12}.$$

Suggestion. — Let $r=\dfrac{s}{6}$ and $h=\dfrac{s}{3}$ and substitute in the formula found in Ex. 5.

7. If $AP=\dfrac{AB}{6}$, find CD so that CD shall be $\frac{2}{7}$ of CO.

$$Ans. \ \frac{AB}{4} \ \text{or} \ -\frac{AB}{18}.$$

Suggestion. — Substitute $r=\dfrac{s}{6}$ in the formula found in Ex. 5. Solve for h the equation $\dfrac{s^2-\dfrac{2\,s^2}{3}+4\,h^2}{8\,h-\dfrac{4\,s}{3}}=\dfrac{7\,h}{2}.$

8. Draw figures showing the meaning of both answers found in Ex. 7.

9. If $CD=\dfrac{AB}{6}$, find AP so that CO shall be $1\frac{5}{8}$ AB.

$$Ans. \ AP=\frac{AB}{8}.$$

342. By varying the data given, other constructions may be obtained for basket-handle arches. In Fig. 287 the arcs which form the arch are AG, GH, and HB with centers at P, O, and Q, respectively. CD < AD.

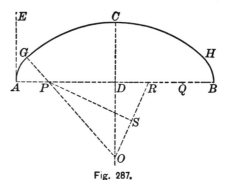

Fig. 287.

1. Given the span AB, the altitude CD, and the point O in the line CD extended, to construct a basket-handle arch with O as center of one of the arcs.

Suggestion. — With O as center and OC as radius, construct the arc GCH. Erect AE perpendicular to AB at A. Construct an arc tangent to AE at A and to the arc GCH. Complete the arch.

2. Complete the circle GCH and show that it is possible to construct two circles tangent to AE at A and to the circle O.

3. Construct the figure according to the suggestions given for Ex. 1 when CO is less than AD. What must be the length of CO in order to form a basket-handle arch?

4. If $AB = s$, $CD = h$, and $CO = r$, find the length of AP.

$$Ans. \ \frac{8\,rh - 4\,h^2 - s^2}{4(2\,r - s)}.$$

Suggestion. — $DO = (r - h)$, $RD = \left(r - \dfrac{s}{2}\right)$. Find OR. From the similar $\triangle\ DRO$ and PSR, $PR = \dfrac{2\,\overline{RS}^2}{DR}$, $AP = (r - PR)$.

5. If $CO = \dfrac{3\,AB}{2}$ and $CD = \dfrac{AB}{6}$, find AP. $\hspace{2em}$ *Ans.* $\dfrac{s}{9}$.

Suggestion. — Substitute $r = \dfrac{3\,s}{2}$ and $h = \dfrac{s}{6}$ in the result obtained in Ex. 4.

6. If $CO = \dfrac{3\,AB}{4}$, find CD so that AP shall equal $\dfrac{5\,s}{18}$.

$$Ans. \ \frac{s}{3} \text{ and } \frac{7\,s}{6}.$$

Suggestion. — Substitute $r = \dfrac{3s}{4}$ in the formula obtained in Ex. 4 and solve for h the equation $\dfrac{5\,s}{18} = \dfrac{6\,sh - 4\,h^2 - s^2}{2\,s}$.

7. Draw a figure illustrating the meaning of both answers obtained in Ex. 6.

8. If $CO = \dfrac{3\,AB}{4}$ and $CD = \dfrac{AB}{6}$, find AP. Draw a figure illustrating the meaning of the result obtained. $\hspace{1em}$ *Ans.* $AP = -\dfrac{s}{18}$.

9. Show that $8\,rh - s^2 - 4\,h^2$ must be positive or no arch of the type here discussed is obtained provided that $r > \dfrac{s}{2}$.

Suggestion. — In the formula $AP = \dfrac{8\,rh - 4\,h^2 - s^2}{8\,r - 4\,s}$, if $8\,rh - 4\,h^2 - s^2$ is negative, a figure is obtained as suggested in Ex. 8.

10. Show that in the figure OA must be less than CO or no arch of the type here discussed is obtained.

11. If $DO = r - h$ and $AD = \dfrac{s}{2}$, find AO and show that Ex. 9 states algebraically the condition that is stated geometrically in Ex. 10.

Suggestion. — $\overline{AO}^2 = \dfrac{s^2}{4} + (r - h)^2$. From Ex. 10 $AO < CO$ or $\dfrac{s^2}{4} + (r - h)^2 < r^2$.

343. The following construction for the basket-handle arch is given in Hanstein, Pl. 22. AB and CD are given.

Construct the equilateral $\triangle\,AED$ on the half span AD. Make DF = DC. Draw CF, meeting AE at G. Draw GP parallel to ED, meeting CD at O. Make BQ = AP. P, O, and Q are the centers, and AP, OC, and QB, the radii for the arcs.

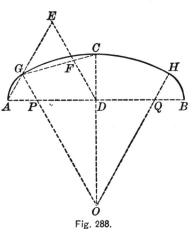

Fig. 288.

EXERCISE

Prove that (1) $OC = OG$, (2) $GP = PA$, (3) $\overset{\frown}{CG}$ and $\overset{\frown}{AG}$ are tangent at G, and (4) $\overset{\frown}{GCH}$ and $\overset{\frown}{HB}$ are tangent at point H on the line OQ.

344. The following construction for the basket-handle arch is given in Hanstein, Pl. 21.   *AB* and *CD* are given.

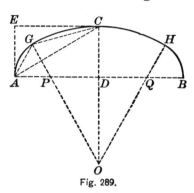

Construct the rectangle ADCE with AD and CD as sides.   Draw AC.   Bisect ∠EAC and ∠ECA.   Let the bisectors meet at G. From G drop a perpendicular to AC; extend the perpendicular to meet CD at O. Make BQ = AP.   P, O, and Q are the centers and AP, OC, and QB the radii for the arcs.

Fig. 289.

EXERCISES

1. Prove that (1) $PA = PG$, (2) $GO = CO$, (3) $\widehat{AG}$ and $\widehat{GC}$ are tangent at point G, and (4) $\widehat{GCH}$ and $\widehat{HB}$ are tangent at point H on the line OQ.

2. Construct Fig. 289 when $CD > \frac{1}{2} AB$.

3. Construct Fig. 288 when $CD > \frac{1}{2} AB$.

4. Study the exercises in §§ 343 and 344, for a figure in which $AP = AD$; for a figure in which $AP > AD$.

5. If AP is equal to or greater than AD, can a basket-handle arch be constructed by methods suggested in §§ 341, 342 ?

TUDOR ARCHES

345. History. — In origin and advantage the Tudor arch is similar to the basket-handle arch.[1]   While the latter is more common in France, the use of the Tudor arch was largely confined to England.   The Tudor arch is seen on many university and college buildings in this country.

[1] Bond, p. 266.

346. The Tudor arch may be constructed if the length of the span and the altitude, the positions of the centers of the two end circles, and the direction of the tangent at the apex are given.[1]

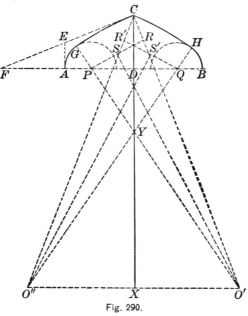

Fig. 290.

In Fig. 290 CD and AB are of fixed length. CD is the perpendicular bisector of AB. P and Q are fixed points so that AP = BQ. CE is a straight line making any arbitrary acute angle with CD.

EXERCISES

1. Construct an arc tangent to *CE* at *C* and to the circle *P*. Show that two such circles are possible.

2. Let *CE* intersect *BA* or *BA* extended at *F*. Construct a figure to illustrate each of the following special cases :

(*a*) *CD* > *AP* and point *A* between *F* and *P*.
(*b*) *CD* > *AP* and point *F* between *A* and *P*.
(*c*) *CD* > *AP* and point *F* between *P* and *D*.
(*d*) *CD* = *AP*.   (*e*) *CD* < *AP*.

---

[1] *Enc. Brit.* : Building.

3. What are the conditions for forming a Tudor arch as shown in Fig. 290?

4. Construct the Tudor arch as shown in Fig. 290, making the half *CDB* symmetrical with *CDA*. Lines *AB*, *CD*, and *CE* and point *P* are given in position.

5. Prove that lines *PR*, *QR'*, and *CD* are concurrent.

6. Prove that line *O'O''* is parallel to line *AB* and is bisected at right angles by *CD* extended.

7. Prove that lines *PO'*, *QO''*, and *CX* are concurrent; also lines *SO'*, *S'O''*, and *CX*.

8. Construct the figure so that *CG* and *CH* are straight lines. Let *P* and *Q* be fixed points.

9. In Fig. 290 let lines *AB*, *CD*, and *CE* be given in position. Construct a perpendicular to *AB* at *A*, intersecting *CE* at *E*. If point *P* is arbitrary, for what positions of point *P* is the arch possible?

10. For what position of point *P* will *CE* be tangent to circle *P*, if lines *CE*, *CD*, and *AB* are fixed in position?

347. The following construction for Tudor arches has been given. It is of limited application as the height of the arch is fixed by the construction.

In Fig. 291 AB is of given length. CD, of indefinite length, is the perpendicular bisector of AB. AP = PD = DQ = QB. ABO'O'' is a square erected on AB. Lines O'P and O''Q are drawn and extended. Points P, O', O'', and Q are the centers for $\widehat{AG}$, $\widehat{GC}$, $\widehat{CH}$, and $\widehat{HB}$, respectively.

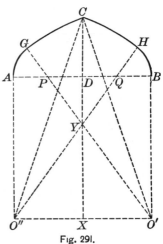

Fig. 291.

1. Show how to construct the complete figure.

2. Prove that $\overset{\frown}{AG}$ and $\overset{\frown}{CG}$ are tangent; also $\overset{\frown}{BH}$ and $\overset{\frown}{HC}$.

3. Prove that $\overset{\frown}{CG}$ and $\overset{\frown}{CH}$ and the line CD are concurrent.

4. If $AB = s$, find the length of GO' and CD.

$$Ans. \text{ (1) } \frac{3\,s}{2}; \text{ (2) } s(\sqrt{2}-1).$$

5. Construct a figure similar to Fig. 291, dividing AB into three equal parts.

6. Find the length of GO' and CD for the figure constructed in Ex. 5.

348. The following construction from Becker, page 80, like the preceding construction, is of limited application.

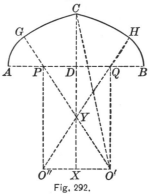

In Fig. 216 AB is of given length. CD, of indefinite length, is the perpendicular bisector of AB. $AP = PD = DQ = QB$. The rectangle $PO''O'Q$ is so constructed that PO'' is $\frac{3}{2}$ of PQ. P, O′, O″Q are the centers for $\overset{\frown}{PG}$, $\overset{\frown}{GC}$, $\overset{\frown}{CH}$, and $\overset{\frown}{HB}$, respectively.

Fig. 292.

1. Show how to construct the complete figure.

2. Prove that $\overset{\frown}{AG}$ and $\overset{\frown}{GC}$ are tangent; also $\overset{\frown}{HB}$ and $\overset{\frown}{HC}$.

3. Prove that $\overset{\frown}{GC}$ and $\overset{\frown}{CH}$ and the line CD are concurrent.

4. If $AB = s$, find the length of GO' and CD.

$$Ans. \text{ (1) } \frac{s}{4}(1 + \sqrt{13}); \text{ (2) } \frac{s}{4}(\sqrt{13 + 2\sqrt{13}} - 3).$$

5. Construct a figure similar to Fig. 292, dividing AB into five equal parts, and making $PO'' = \frac{4}{3} PQ$ and $PQ = \frac{3}{5} AB$.

349. Becker, page 80, gives also the following special construction for a Tudor Arch.

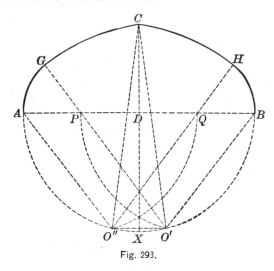

Fig. 293.

In Fig. 293 AB is of given length.   AP = PD = DQ = QB.   AO″O′B is a semicircle constructed on AB as diameter.   P and Q are the centers and PQ the radius for $\widehat{PO'}$ and $\widehat{QO''}$.   With P as center and AP as radius draw arc AG; with O′ as center and O′G as radius draw arc CG.   Similarly Q and O″ are centers for $\widehat{BH}$ and $\widehat{HC}$.

EXERCISES

1. Prove that $PO' = O'B = AO'' = QO''$.

2. Prove that $\widehat{AG}$ and $\widehat{GC}$ are tangent; also $\widehat{BH}$ and $\widehat{CH}$.

3. Draw CDX, the perpendicular bisector of AB.   Prove that $O'O''$ is parallel to AB and equal to $\frac{1}{4} AB$, and that CX is the perpendicular bisector of $O'O''$.

4. Prove that the arcs GC and CH and the line CD are concurrent.

5. If $AB = s$, find the altitude of $\triangle AO''Q$.   *Ans.* $\frac{s}{8} \sqrt{15}$.

Suggestion. — Draw $O''D$ which equals $\frac{s}{2}$.   Draw altitude from O'' meeting AB at K,   Use $\triangle KO''D$.

BIBLIOGRAPHY

History of Math

Aaboe, Asger. *Episodes From the Early History of Mathematics.* New York: L. W. Singer, 1964.

Ball, W. W. Rouse. *A Short Account of the History of Mathematics.* New York: Dover Publications, 1908.

Barrios-Chacon, Miriam, et al. *Multiculturalism in Mathematics, Science, and Technology: Readings and Activities.* Menlo Park, CA: Addison-Wesley, 1993.

Bergamini, David. *Mathematics.* New York: Time, 1963.

Coolidge, Julian. *The Mathematics of Great Amateurs.* New York: Dover Publications, 1963.

Edeen, Susan, John Edeen, and Virginia Slachman. *Portraits for Classroom Bulletin Boards: Mathematicians, Book 1.* Palo Alto, CA: Dale Seymour Publications, 1990.

Edeen, Susan, John Edeen, and Virginia Slachman. *Portraits for Classroom Bulletin Boards: Mathematicians, Book 2.* Palo Alto, CA: Dale Seymour Publications, 1990.

Edeen, Susan, John Edeen, and Virginia Slachman. *Portraits for Classsroom Bulletin Boards: Women Mathematicians.* Palo Alto, CA: Dale Seymour Publications, 1990.

Egsgard, John. *Making Connections with Mathematics.* Providence, RI: Janson, 1988.

Grinstein, Louise, and Paul Campbell, eds. *Women and Mathematics: A Biobibliographic Sourcebook.* Westport, CT: Greenwood, 1987.

Historical Topics for the Math Classroom. Reston, VI: National Council for Teachers of Mathematics, 1989.

Johnson, Art. *Classic Math: History Topics for the Classroom.* Palo Alto, CA: Dale Seymour Publications, 1994.

Newman, James R., ed. *The World of Mathematics. Vol. I, II, III, IV.* Redmond, WA: Microsoft, 1988.

Osen, Lynn. *Women in Mathematics.* Cambridge, MA: MIT, 1974.

Pappas, Theoni. *Mathematics Appreciation.* San Carlos, CA: Math Aids, 1986.

Perl, Teri, and Joan Manning. *Women, Numbers, and Dreams.* Santa Rosa, CA: National Women's History Project, 1985.

Perl, Teri. *Math Equals.* Menlo Park, CA: Addison-Wesley, 1978.

Perl, Teri. *Women and Numbers.* San Carlos, CA: World Wide Publishing, 1993.

Reid, Constance. *A Long Way from Euclid*. New York: T.Y. Crowell, 1963.

Reimer, Luetta and Wilbert. *Mathematicians Are People, Too*. Menlo Park, CA: Dale Seymour Publications, 1990.

Smith, D. E. *History of Mathematics. Vol. I, II*. New York: Dover Publications, 1951.

Struik, Dirk J. *A Concise History of Mathematics*. 4th ed. rev. New York: Dover Publications, 1987.

Turnbull, Herbert Westren. *The Great Mathematicians*. New York: New York University Press, 1961.

Zaslavsky, Claudia. *Africa Counts: Number and Pattern in African Culture*. New York: Lawrence Hill Books, 1990.

History of Geometry

Beckman, Petr. *A History of π (Pi)*. 5th ed. New York: St. Martin's Press, 1971.

DeLacy, Estelle A. *Euclid and Geometry*. New York: Watts, 1963.

Diggins, Julia E. *String, Straightedge, and Shadow: The Story of Geometry*. New York: Viking Press, 1965.

Dunham, William. *Journey Through Genius: The Great Theorems of Mathematics*. New York: Wiley, 1990.

El-Said, Issam and Ayse Parman. *Geometric Concepts in Islamic Art*. Palo Alto, CA: Dale Seymour Publications, 1976.

Grunbaum, Branko and G. C. Shephard. *Tilings and Patterns: An Introduction*. New York: W.H. Freeman, 1989.

Kappraff, Jay. *Connections: The Geometric Bridge Between Art and Science*. New York: McGraw-Hill, 1991.

Kolpas, Sidney J. *The Pythagorean Theorem*. Palo Alto, CA: Dale Seymour Publications, 1992.

Schattschneider, Doris. *Visions of Symmetry: Notebooks, Periodic Drawings, and Related Work of M.C. Escher*. New York: W. H. Freeman, 1990.

Geometric Design

Bezuska, Stanley, Margaret Kenney, and Linda Silvey. *Designs from Mathematical Patterns*. Palo Alto, CA: Dale Seymour Publications, 1990.

Blackwell, William. *Geometry in Architecture*. Berkeley, CA: Key Curriculum Press, 1984.

Bourgoin, J. *Arabic Geometrical Pattern and Design*. New York: Dover Publications, 1973.

Hilbert, David. *Geometry and the Imagination*. New York: Chelsea, 1952.

Horemis, Spyros. *Optical and Geometrical Patterns and Designs.* New York: Dover Publications, 1970.

Millington, Jon. *Curve Stitching.* Norfolk, England: Tarquin Publications, 1989.

Oliver, June. *Polysymetrics.* New York: Parkwest Publications, 1986.

Pohl, Victoria. *How to Enrich Geometry Using String Designs.* Reston, VI: National Council for Teachers of Mathematics, 1986.

Posamentier, Alfred S. and William Wernick. *Advanced Geometric Constructions.* Palo Alto, CA: Dale Seymour Publications, 1988.

Row, T. Sundara. *Geometric Exercises in Paper Folding.* New York: Dover Publications, 1966.

Seymour, Dale and Jill Britton. *Introduction to Tessellations.* Palo Alto, CA: Dale Seymour Publications, 1989.

Seymour, Dale. *Geometric Design.* Palo Alto, CA: Dale Seymour Publications, 1988.

INDEX TO PROBLEMS AND THEOREMS

Numbers refer to paragraphs and exercises; thus, 94 (10) means § 94, Ex. 10.

Circles, construction of:

INDEX TO PROBLEMS AND THEOREMS

b. Lengths of lines : 59 (3–9).

c. Determination of ratios : 51 (2, 4) ; 59 (3–9) ; 61 (2, 3) ; 62 (4, 5).

III. Ratios derived from given data, used to form parallel lines : 51 (1) ; 54 (1 3) ; 81 (1) ; 98 (1) ; 129 (1) ; 131 (2) ; 142 (1) ; 143 (1) (see suggestion) ; 316 (1) ; 317 (1) ; 320 (1).

IV. Similar triangles, used to find lengths of lines : 59 (4–9) ; 69 (3,5) ; 70 (6) ; 275 (20, 21) ; 301 (13–15) ; 303 (5) ; 315 (2) ; 319 (3) ; 334 (4, 5, 8) ; 335 (3,6) ; 341 (4, 5) ; 342 (4).

Proportionals, construction of :

I. Fourth proportional : 66 (15).

II. Mean proportional : 253 (1).

III. Proportional segments : 54 (14) ; 56 (5, 7) ; 71 (17, 20) ; 84 (3) ; 96 (4) ; 272 (10) ; 310 (2) ; 311 (3).

Pythagorean theorem :

I. Proof for : 59 (12).

II. Used for :

a. Lengths of lines. (See *Lines, lengths of.*) Harder problems : 258 (2) ; 259 (2) ; 260 (3) ; 262 (3, 4) ; 263 (3, 4).

b. Formation of equations. (See *Equations linear, I,* and *Equations quadratic, I, containing binomial squares.*)

Quadratics. (See *Equations, quadratic.*)

Quadrilaterals :

I. **Areas of :** 74 (2) ; 88 (5) ; 98 (4, 9).
(See *Parallelograms, Rectangles, Rhombuses, Squares, Trapezoids.*)

II. **Congruence of :** 13 (5) ; 63 (2) ; 67 (1) ; 69 (2) ; 70 (4) ; 72 (1) ; 73 (1) ; 74 (1) ; 150 (10) ; 151 (7, 8) ; 154 (3).

Radical equations. (See *Equations.*)

Radicals. (See *Square roots.*)

Ratios of areas, involving :

I. Arithmetical computations, figures derived from :

a. Parallelograms and triangles : 44 (4) ; 45 (8) ; 102 (6).

INDEX TO NOTES AND ILLUSTRATIONS

Numbers in Roman type refer to paragraphs and exercises, as in the previous index. Numbers in italics refer to illustrations.

INDEX TO REFERENCES

Numbers refer to pages.

INDEX TO REFERENCES